PORT MOODY PUBLIC LIBRARY

D0092650

Praise for *Superbug*

"A wonderful book."

—Dr. Michael Roizen, coauthor of *You: The Owner's Manual*
and contributor to HealthRadio.net

"Lays bare, often all too graphically, the ravages of a disease with the potential to do grievous international harm."

—*Booklist*

"If you want the short version, *Superbug: The Fatal Menace of MRSA* is a must-read. . . . McKenna makes it clear that this epidemic arose and persists because of human error."

—*ScienceBlogs*

"[McKenna] examine[s] the world of viral and bacterial killers and the organization—the Centers for Disease Control and Prevention—that attempts to keep microscopic monstrosities in check. . . . [She] looks at the outbreak and rise of MRSA itself, and what lies ahead for us."

—*The Atlanta Journal-Constitution*

"This is the first book on MRSA, and it's scary."

—*The Orange County Register*

Praise for *Superbug*

"A wonderful book."
—Dr. Michael Rosen, coauthor of *Baru, the Queen's Maiden* and contributor to the ...

"They are ... often all too graphically the ravages of a disease with the potential to be a serious international harm."
—*Nature*

"Of pandemic short version, SuperBug ... The ... First Advance of MRSA has much to ... McKenna makes it clear that this epidemic arose and persists because of human error."
—*Kirkus Reviews*

"McKenna's examples ... the waste of total and bacterial killers and the ingratitude—the Centers for Disease Control and Prevention—that struggles to keep antiseptic resistance in check. ... This book is at the outbreak and rise of MRSA itself, and what lies ahead for us."
—*The Atlanta Journal-Constitution*

"Here's the first book on ... McKenna's story."
—*The Orange County Register*

SUPERBUG

The Fatal Menace
of MRSA

MARYN McKENNA

FREE PRESS
New York   London   Toronto   Sydney

FREE PRESS
A Division of Simon & Schuster, Inc.
1230 Avenue of the Americas
New York, NY 10020

Copyright © 2010 by Maryn McKenna

All rights reserved, including the right to reproduce this book or portions thereof
in any form whatsoever. For information address Free Press Subsidiary Rights Department,
1230 Avenue of the Americas, New York, NY 10020

First Free Press trade paperback edition February 2011

FREE PRESS and colophon are trademarks of Simon & Schuster, Inc.

For information about special discounts for bulk purchases,
please contact Simon & Schuster Special Sales at 1-866-506-1949
or business@simonandschuster.com.

The Simon & Schuster Speakers Bureau can bring authors to your live event.
For more information or to book an event contact the Simon & Schuster Speakers Bureau
at 1-866-248-3049 or visit our website at www.simonspeakers.com.

Manufactured in the United States of America

10   9   8   7   6   5   4   3   2   1

The Library of Congress has cataloged the hardcover edition as follows:

McKenna, Maryn.
Superbug : the fatal menace of MRSA / Maryn McKenna—1st Free Press hardcover ed.
p. cm.
Includes bibliographical references and index.
1. Staphylococcus aureus infections. 2. Methicillin resistance.
I. Title. [DNLM: 1. Methicillin-Resistant Staphylococcus aureus—United States.
2. Health Policy—United States. 3. History, 20th Century—United States.
4. History, 21st Century—United States. WC 250 M478s 2010]
QR201.S68M45   2010
616.9'297—dc22   2009037793

ISBN 978-1-4165-5727-2
ISBN 978-1-4165-5728-9 (pbk)
ISBN 978-1-4391-7183-7 (ebook)

For Loren

Everybody knows that pestilences have a way of recurring in the world; yet somehow we find it hard to believe in ones that crash down on our heads from a blue sky. There have been as many plagues as wars in history; yet always plagues and wars take people equally by surprise.

—ALBERT CAMUS, *The Plague*

CONTENTS

AUTHOR'S NOTE

I first became aware of methicillin-resistant *Staphylococcus aureus*, MRSA, in 2003, when I shadowed a group of epidemiologists from the U.S. Centers for Disease Control and Prevention and the Los Angeles Department of Public Health through an outbreak investigation. Disease detection was not new to me: I was the CDC beat reporter for the *Atlanta Journal-Constitution*, and my job was to chase the CDC as it chased epidemics around the world. My newsroom colleagues jokingly called me Scary Disease Girl.

The Los Angeles outbreak, though, was not so much frightening as it was puzzling. Serious MRSA skin infections were appearing in gay men who frequented sex clubs. The task facing the disease detectives was finding out whether the bug was behaving as it always had, passing between people who had contact with each other's skin, or with surfaces such as benches that skin had recently touched, or whether it had taken on the characteristics of a sexually transmitted disease. It was an intriguing outbreak, full of questions about policy and publicity, sexual behavior and bacterial unpredictability, and I marked it down mentally as interesting and moved on. (The verdict, in the end: skin-to-skin and skin-to-surface contact, no different from a gym.)

Fast-forward three years, and I had left my newspaper job and was studying for twelve months in the media fellowship program of the Henry J. Kaiser Family Foundation, investigating overcrowding and stress in emergency rooms. I spent eight overnight shifts a month in ERs in several cities. In every city, MRSA was everywhere I looked. Patients had minor skin infections, huge gaping wounds, grave infestations of bone and muscle, and critical pneumonias that got them whisked immediately to the ICU. I cannot forget one man who unbuttoned the cuffs of a

xi

crisply creased dress shirt to reveal that he had wrapped both forearms in hand-towels—because bandages were not sufficient to contain the seepage from infections that had eaten through his skin and deep into muscle, exposing the white flash of tendons in the open pit of his elbow. I was horrified. This was much more serious than boils from a gym bench. But the attending physician I was shadowing that night, though compassionate, was otherwise unimpressed, because he saw some form of MRSA infection, usually many MRSA infections, on most nights of the week.

A year later, I wrote a story for *Self* magazine about the unappreciated threat that community-associated MRSA posed to women and children—unappreciated because, at the time, coverage of MRSA in most of the press was limited to reports of outbreaks in prisoners and sports teams, which is to say, men. If I needed a measure of how much MRSA's incidence had changed since my first encounter with the bug, the reaction to that 2007 story gave it to me. Almost one hundred women, and a number of men as well, emailed me, called me, or wrote me anguished letters about how MRSA had disrupted their families and drained their bank accounts. Shortly afterward, I began publishing a blog to gather MRSA stories, also called *Superbug*. It has been running for more than two years now, and in that time, not a week has gone by but someone has contacted me unprompted because they need to describe how MRSA has shattered their lives.

For most of my career as a reporter, I took pleasure in being a disease geek talking to other disease geeks, telling stories that combined the surprises of biology and the intricacies of epidemiology with a faint frisson of fear. But when I examined my professional life, I realized that many of the outbreaks I drew my readers into were caused by diseases that may have fascinated them, but did not imminently threaten them. When I contemplated how MRSA advanced while I was not paying attention, I felt as though I had been telling my readers to stare up at the mountains for fear one of them might one day fall on us, while all the time a flood was rising unnoticed around our feet.

MRSA is that flood. It is a crisis in many dimensions. Its advance exposes huge gulfs between medical researchers and health care workers, between health care and public health, between public health and the public it is intended to serve. It illustrates failures of science, failures of the marketplace, and failures of support for research and innovation. It touches the way we fund our schools and schedule our work lives, and it demonstrates that the way we raise food animals is not sustainable or safe.

AUTHOR'S NOTE

As *Superbug* shows, the history of MRSA is the history of three overlapping epidemics: in hospitals, in everyday life, and now in animals. We could have responded to each as they occurred—when our toes were wet, or our ankles. We failed. Now the flood is somewhere around our knees, and rising.

Maryn McKenna
Minneapolis, 2009

SUPERBUG

SUPERBUG

THE FIRST ALERT

Tony Love's knee ached.

The rangy, round-headed thirteen-year-old had banged into a friend a week ago while they were playing volleyball in the school gym. They crashed to the floor together, arms and untied shoelaces flying, and Tony scraped his elbow. After school, he and his mother and his grandmother had bandaged the cut and shrugged it off. He was a teenager, after all; Clarissa Love, his mother, expected her son to be rambunctious. It was mid-September 2007. The weather was still hot south of Chicago and Tony was still in summer mode, twitching behind his desk at school until the bell rang and he could burst out and work it off. The scratch was no big deal, and Tony was tough; he was the second child of six, and the only boy until his baby brother, the youngest, had come along. Tony saw himself as the man of the family, keeping his sisters in line while Clarissa, who was thirty, worked as an aide for the disabled.

The elbow had healed up after a few days, but then his left knee started to hurt. Now it was hot and so swollen he couldn't bend his leg. When he tried to put his weight on it, it throbbed like his heart had gone down behind his kneecap. Clarissa had gone away for a few days, so her mother Sandra put the oldest sister in charge of the other children, hooked Tony's arm around her shoulder, and steered him out to the car. He leaned on her heavily, hopping on his good leg and wincing when the other foot hit the ground.

At the little local hospital, the emergency room doctor listened to Tony's story and shrugged. It was probably a sprain, he said; take the boy home, give him Motrin, wrap the knee in hot towels, and it would be better in a few days. They staggered home.

It did not get better. Four days later, Tony's left knee still hurt, and his left foot and both of his hands did too. His hip joints ached so much he didn't want to walk, not even to the bathroom. He didn't want to eat, either. A thirteen-year-old boy with no appetite; to his grandmother, that was the biggest warning sign of all. She checked his temperature and found it was 104. Frightened, she hauled him out to the car and took him to the next-biggest local hospital, a few miles further south. The ER staff there checked his vital signs and listened to his story: the scrape, the fever, the lethargy, the joint pain for more than a week, the not wanting to eat or pee.

They were a little worried, they told his grandmother. Tony's pulse and blood pressure looked normal and his breathing was fine, but the fever indicated an infection, and his kidneys weren't working as well as they should. The hospital was willing to admit him, but to be safe, the ER staff thought they ought to take him to a children's hospital. There was a very good one, they said, back toward the city, at the University of Chicago, and they called an ambulance.[1]

It was the end of the workday, and Clarissa met Tony and her mother at Comer Children's Hospital, a gleaming new glass pile just off the university's park-like main boulevard. The ambulance crew that brought them rolled Tony straight up to the medical floor, and the nursing staff began admitting him, checking his vital signs again and going over his paperwork from the smaller hospital. The ER staff there had suspected that Tony had osteomyelitis, a bone infection that could be caused by several kinds of bacteria. It was a serious condition, but not rare, and it was treatable, requiring that he get the right drugs for whichever bacteria were infecting him and be monitored by someone who understood the disease in children.

But while they were talking, Tony's condition abruptly got worse. He became agitated and confused; then he began breathing fast and deep. His skin had been radiating heat from the fever, but it turned cold as quickly as if someone had parked him in front of an air conditioner. The medical staff around him recognized the signs: the bacterial infection was spilling over into his bloodstream, and his immune system's spiraling reaction was slowing his pulse and crashing his blood pressure. In half an hour, he had gone from a sick kid to a kid in crisis.

A nurse phoned urgently upstairs to the pediatric intensive care unit, checking for an open bed that had all the monitoring equipment they would need. The technicians kicked the gurney's brake locks and got him rolling, skidding past the curvy computer stations and the kid-friendly bright red columns. Tony was sliding into septic shock, and that was an emergency. Inside his body, chemicals released by his immune system were triggering a cascade like dominos falling. They were stretching the firm walls of his blood vessels, making them porous, and fluid was leaking out into his tissues. Blood cells were clumping and clogging his capillaries, and his oxygen-starved organs were beginning to fail. Clarissa felt her stomach cramp in fear. In front of her eyes, her son was dying.

In the ICU, the staff sedated Tony and slid a tube down his throat, turning the hard work of breathing over to a ventilator. They threaded IVs into his veins and hooked him to bags of fluids, plugging in four drugs to bring back his blood pressure and stimulate and stabilize his heart rate, and four more drugs to contain whatever bacteria were revving his immune system into overdrive.

To his bewildered mother and grandmother, the swirl of controlled chaos around Tony was as inexplicable as his sudden collapse; the ICU staff seemed to be trying everything, hoping it would bring him back from the brink. No diagnosis was possible yet. They had been in the hospital barely an hour, not long enough for test results to make it down to the lab and back. But the medical staff had a strong suspicion of what could bring a healthy boy down so quickly, and the clue lay in one of the drugs they ordered pushed into his veins. It was called vancomycin, and it was famous in hospitals as a drug of last resort. They used it against a bacterium that had learned to protect itself against most of the other drugs thrown at it, a particularly dangerous variety of staph called methicillin-resistant *Staphylococcus aureus*—MRSA, for short.

Staph, the short form of the family name *Staphylococcus*, is an ancient organism with a vast arsenal of tricks and defenses, some of them newly learned, others as old as man. It is unpredictable, dynamic, potentially deadly—and for more than a decade, it had been the obsession of a small group of University of Chicago researchers. Geographic accident had brought Tony to a place that understood how to help him, but it was far too soon to know whether he had arrived in time.

Orthopedic surgeons and plastic surgeons converged on the room Tony had been hastily stashed in. The fever, the septic shock, the pain in his legs and joints—all the symptoms indicated the infection was making abscesses that would need to be opened and drained immediately. The teams ran him quickly through radiology for a CT scan, peering at the screen for the bright white spots that indicate infection, and then to the operating room to get him prepped and anesthetized.

Plastic surgeons are the watchmakers of medicine, practiced at maneuvering in tight areas packed with crucial interconnected parts. They went to work on Tony's left hand, cutting carefully through ligaments and tendons to preserve as much function as possible. Inside his fingers, they found pockets of pus the size of nickels. There was one in the center of his hand; it was the size of a golf ball. There were others in his right hand, too, and more hidden beneath the bones of his right foot. Orthopedic surgeons are cabinetmakers, trusted to protect the strength of the body's scaffolding and the smooth function of its joints. They probed Tony's hips and shoulders with a long wide-bore needle, looking for infection trapped behind the joints' cartilaginous sheaths. His left knee, the one he couldn't bend, was rigid and swollen. When they slid the needle in, pus pushed out under pressure, forcing back the base of the syringe. They got out enough to fill a baseball.

One of the orthopedic surgeons sliced into Tony's left thigh and eased apart the muscles. There was pus underneath them, creamy and dull. There was too much to evacuate through the small incision they had cut, so they kept cutting, looking for the end of the pocket. They laid his thigh open from his knee almost to his hip joint; wherever they cut, they found a dense deposit of pus wrapped around the bone. They used a tool like a dentist's jet to work it free, rinsing the cavity between bone and muscle with high-pressure water and sucking the slurry away. The abscess was so deep that they could not trust they had cleaned out all the infection, and so they left the gash open. They wrapped it in dressings that would let the mess drain, and rolled him back to the ICU.

They brought Tony back to a room at the center of the unit, as close as they could put him to the nurses who would monitor his every moment. He was still sedated and intubated and teetering on the verge of shock. He had pneumonia, and his liver was not clearing waste products from his blood. The intensive-care team pumped him with drugs and fluids: antibiotics to kill the still-unidentified bacteria, immune globulin to neutralize toxins, vasopressors to keep his blood pressure up. The drug doses had to be maintained in a delicate, shifting balance. Too much or too little could send his heart into an off-kilter rhythm, or scatter small clots through his bloodstream, or clamp down the small vessels in his extremities and kill a finger or toe.

Not long after Tony came back to the ICU, the unit's computer pinged with the first report from the hospital's microbiology lab. The results validated the intuition of the health care workers who had ordered him onto vancomycin many hours earlier. Tony did have MRSA.

4

"They told me he was the sickest child on that ICU," Clarissa recalled. "They didn't expect him to live."

The Chicago group's long journey with MRSA began in 1996, eleven years before Tony rolled through their emergency room doors. It started with a casual hallway conversation between Dr. Robert Daum, the chief of pediatric infectious diseases, and Dr. Betsy Herold, a faculty physician. They had both noticed that they were suddenly seeing a lot more kids with staph.[2]

That staph infections were occurring was not, in itself, remarkable. *Staphylococcus* is a large genus of bacteria, and *S. aureus*—the strain that is most common in humans, and that most people mean when they say "staph"—is probably one of mankind's oldest evolutionary companions. Over millions of years, it has learned to live benignly on human skin and in human nostrils, in a microscopic intimacy that biologists call "commensal," from the Latin words for "being at table together." At any moment, one out of every three of us is carrying S. aureus without being made sick by it.[3] But when that mutual balance is disrupted, S. aureus can attack with ferocity, causing infections that range from simple skin boils and rashes to muscle and bone abscesses, pneumonia, toxic shock, even infestations of the valves of the heart.[4] The countless generations of close contact have given the bacterium an unmatched familiarity with the human immune system, and out of that long acquaintance it has evolved a huge range of microbiological weapons called virulence factors—more than seventy cell-destroying enzymes and toxins, many more than any other bacterium can produce.[5]

Staph is a fearsome aggressor when the body's protective mechanisms are disrupted, both the first-line defense of our skin and the complex chemical weaponry of the immune system. It is notorious for invading not only large surgical wounds, but the small incisions made to accommodate IV lines and dialysis catheters; it enters the body through the smallest gap, and it forms sticky, infectious films on the tubes passing through those gaps. It is a grave danger to anyone whose immune defenses have been reduced by age, illness, or treatment for disease: newborn infants, chemotherapy recipients, people whose poor kidney function forces them into regular dialysis. It is by far the most common cause of infections in hospital patients, causing almost a half-million serious illnesses every year.[6]

And in addition to its virulence, staph possesses another potent weapon: it has been developing successful defenses against antibiotics for as long as antibiotics have existed. That is not a long time, since penicillin only went into wide use at the end of World War II. But staph is so adept at evading threats to its existence that it learned to protect itself against penicillin almost as soon as the earliest experimental doses of the drug were deployed, and long before it was placed on the open market.[7] Ever since, bug and drugs had been locked in a lethal game of leapfrog, with staph always one leap ahead. After penicillin, pharmaceutical chemists created a new drug, methicillin, with a chemical structure that had never before existed in nature—a strategy they thought would provide decades of protection against staph's adaptive brilliance. But methicillin, launched in 1960, was not even on the market a year before staph developed resistance against it as well.[8] For decades, pharmacologists spun new compounds out of methicillin's structure, hoping to find the perfect molecular shape to penetrate staph's defenses, but they had never been successful for long.*

By the time Daum and Herold paused in the hallway of Wyler Children's Hospital in 1996, methicillin resistance was so widespread that almost every large hospital in the United States had detected at least one MRSA infection, and a third of all the staph cases that occurred during hospital stays were caused by MRSA instead of common drug-sensitive staph.[9] Almost all those MRSA infections had a common history: they began after the victim was admitted to a hospital. The victims were nursing-home residents or cancer patients or had long-term medical problems. They were vulnerable because they were diabetic and took dialysis three times a week, or because they had cystic fibrosis and their lungs were full of sticky mucus that made a great breeding ground for germs.

The cases that Daum and Herold were noticing were not like those patients. They were not old people or babies or anyone who had been sick for a long time. Instead, they were children who had not been in any hospital for as long as their families could remember, but who nevertheless had been felled by an inexplicable bacterial attack.

In twenty-four years in medicine, Daum had seen a lot of staph, and he had developed a rueful respect for the bug's persistence and resilience. He knew better than to assign human qualities to a bacterium, and yet he would slip and speak as if it knew what it was doing. "It knows how to live on inanimate objects, in our noses, in our skin, in our genitals," he

*For a discussion of drugs, drug families, and resistance factors, see the appendix.

would say. "It's virulent, it's adaptive, and it is able to circumvent almost everything we throw at it. It is a perfect pathogen."

By 1996, Daum had been at University of Chicago for eight years. He was a Boston native, but he had stepped aside from the straight-to-Harvard path expected of aspiring doctors in Boston and gone to Montreal for college instead. He stayed for medical school and stayed again for residency—and then succumbed to tribal custom and headed to Harvard at last for a postgraduate fellowship. He still spoke good French, and at home he often listened to old Edith Piaf recordings. Her defiant, luxuriant melancholy was a paradoxical comfort after long days confronting the terrible damage that childhood diseases could do.

The cases he and Herold had noticed stood out not just because they were mysterious but because they were severe. One fourteen-year-old had a raging infection that filled the spongy bone of his heel with pus. Another, only six, had a painful, walled-off abscess, hard with clotted blood, buried deep in the muscle of his butt.[10] The illnesses were as serious as the infections that people contracted if they had been in the hospital, and taken major antibiotics, long enough to undermine their immune systems. But these children had not been in any hospital; nor did they suffer from any of the chronic conditions such as diabetes that could have made them vulnerable. If they had staph at all, it should have been only a rash or an eyelid stye, a minor problem that a kid might pick up in everyday life.

And incontrovertibly, they did have staph. Again and again when the team sucked tiny samples of pus out of the infections and sent them to the hospital's lab, the samples grew into S. aureus's streaky yellow colonies. (*Aureus* comes from the Latin for "golden.") Even more troubling, this staph was drug-resistant. When the lab technicians dropped minuscule dots of drug-impregnated paper onto the staph colonies and watched for the ring-shaped dead zone that would indicate the bug was susceptible to the drug, again and again no ring formed. The children had MRSA—and they had acquired it out in the community, a setting where MRSA was not supposed to be.

A virulent, drug-resistant hospital bug in children who had never been in a hospital? It made no sense. Herold and Daum responded in the classic manner of scientists. They performed a study. Delving into the hospital's archives, they pulled the medical charts of every child hospitalized with S. aureus from mid-1993 to mid-1995, and for comparison's sake pulled charts from five years earlier. They sorted through the sets from the 1980s and the 1990s, dividing them into progressively smaller subsets. First they separated out children with methicillin-susceptible staph, leaving only children with MRSA. Then they separated out children who developed MRSA while they were hospitalized, leaving only children who would have come into the hospital from the community already infected. Finally, they took a second look at the records of the children with those community infections, and separated out any child who had been hospitalized in the six months before the infection, had been intubated or catheterized even as an outpatient, had undergone surgery, or had anyone in their household to which the same conditions applied—anyone, in other words, who had some intimate health care contact that might have passed the bacterium along.

That left them, in each cohort, with some number of children who had developed MRSA despite having had no contact with the health care institutions where MRSA was known to flourish. The researchers compared the two time groups and were shocked: In the past two years, there had been twenty-five cases of MRSA infection that had no link at all to health care. Five years earlier, there had been one.

The medical records revealed that these infections were different from the norm in several striking ways. Over more than three decades, bit by bit, the MRSA that patients contracted in hospitals had developed resistance to an enormous range of drugs. It could disarm not only methicillin, but a vast class of similar drugs, called beta-lactams for a feature of their chemical structure. It had accumulated additional defenses against other drug classes, so many of them that the bacterium was protected against most of the antibiotics that physicians use every day.[11] It could reliably be treated only by one drug, vancomycin. By 1996, when doctors diagnosed MRSA they expected it to be multidrug-resistant. But the MRSA in the Chicago children did not possess multiple resistance. It had defenses against methicillin and the other beta-lactams, but that was all.

In another oddity, the bug in the children with the community infections seemed to have a different genetic structure than the MRSA norm. Because they had seen the children so recently, the Daum team still had samples of their blood stashed in the hospital's laboratory freezers. And because biomedical researchers are pack-rats who never throw away anything that might one day be useful, they also had in the freezer some samples from children who had developed MRSA infections in the hospital sometime in the 1990s. The lab technicians isolated the bacterial DNA in the samples and used a technique called pulsed-field gel electrophoresis—PFGE, for short—to generate a bar code–like pattern that served as each bacterium's molecular fingerprint. When the test was done, the researchers noticed something striking. The bar codes representing the fourteen-year-old and the six-year-old boys—who were not related to each other, and who came to the hospital 6 months apart—were identical to each other, and different from all the rest.

The study confirmed what Herold and Daum had suspected: this strain of MRSA that had walked in from the surrounding neighborhood was different from the MRSA they were used to in perplexing and potentially alarming ways. They wrote an article describing their findings and submitted it to the *Journal of the American Medical Association*. The journal kicked it back.

"They said we didn't know what MRSA strains looked like," Daum said. "So I called up the editor and said, 'You know, I think that's wrong. I think we do know.' And in the end we had to send the genetic evidence that showed incontrovertibly they were MRSA."

The journal eventually published the paper, on February 25, 1998, almost two years after Daum and Herold first discussed their worries. But just a few pages away, it also ran an editorial that suggested the team had just not done their homework.[12] Somewhere in the patients' past, it said, there must be a link to health care—a hospital admission, a clinic visit, a grandparent in a nursing home, or an uncle who was a janitor—that the researchers had simply missed.

For Daum, the medical journal's implicit dismissal of their work was doubly frustrating. The University of Chicago is a private institution where faculty fund their research by means of the grants they attract. The journal had published the work, but did not endorse it; and without the blessing of the medical establishment, funding would be hard to come by. It was exasperating to have caught the scent of something and not have the resources to pursue it.

In the months that followed the paper's publication, other researchers—including disease detectives at the Centers for Disease Control—quietly called and wrote Daum to express similar doubts. That a virulent staph strain could be flourishing outside hospitals, attacking healthy children with no immune-system deficits, challenged everything medicine had discovered about the behavior of drug-resistant organisms. Common opinion agreed with the *JAMA* editorial: the Chicago group had mistakenly latched onto a hospital strain of MRSA that had leaked into the community and persisted there for a while before dying out.

The questioners had some justification for their doubts because something like what they imagined had happened once before. In 1980, a strain of MRSA had spread among heroin addicts in Detroit. There was no MRSA in the community yet, and very little in the city's hospitals; it caused about 7 percent of the infections that occurred in Henry Ford Hospital, the major public institution. Then, over a year and a half, 100 drug addicts living in the city developed MRSA, most with minor problems but about a third with potentially fatal infections of the bloodstream and heart. None of them were being treated in the hospital when the outbreak started, but they were nevertheless almost as medically fragile as hospital patients: undernourished, underexercised, with immune systems already overtaxed. And like hospital patients, they were also multimedicated—not only with heroin or whatever else they could afford, but with black-market antibiotics that they took on the sly to prevent skin infections when they shot up. Many of them had been in and out of Henry Ford's emergency room in the past with minor illnesses, and twelve of them had been admitted to the hospital some time in the previous year. No one else outside the hospital picked up MRSA from the addicts—but one of them carried the strain back into the hospital, where it caused sixty-five more cases, including two nurses and a surgeon who picked it up but were never made sick by it, and fifteen hospital patients to whom the nurses and surgeon unwittingly passed the bug.[13]

Most of the infections in the outbreak began outside the hospital, yet in the view of epidemiologists, it was still a hospital outbreak. The assumption was that one of the addicts (no one could ever say who) had been treated at Henry Ford, picked up MRSA there, carried it outside, and passed it on. Limited lab evidence supported that view. PFGE, the molecular-fingerprint technique, had not been invented yet,[14] but the MRSA strains in the addicts and in the Henry Ford patients were resistant to exactly the same medications: methicillin and all the beta-lactams, and two additional classes of drugs.

By the time the Chicago group's paper was published sixteen years later, most MRSA researchers in the United States were aware of the Detroit outbreak and in agreement about what they thought it represented: the bug could briefly escape from hospitals via discharged patients, but it could not survive long on the outside because healthy people's immune systems defended against it too well. To people who wanted to believe the Chicago team had made a mistake, the Detroit outbreak made a perfect counterargument. They did not consider that over

10

those sixteen years, enough time for millions of bacterial generations to be born, breed, and die, staph had been doing what it does best: evolving.

There were a few clues, buried deep in the medical literature, that something about MRSA's behavior was changing. In 1986, New York City physicians reported at a conference that they had found a group of drug users in Harlem infected with a MRSA strain that did not match the strain in local hospitals because it was resistant to fewer drugs.[15] Then, in 1993, one of those physicians wrote a letter to a medical journal about another strange MRSA case. He had seen a sixty-five-year-old woman from the Bronx who had a staph infection of the heart so severe that it caused a stroke, leaving her paralyzed, deaf, and unable to speak.[16] Her case was a tragedy, but it was also a puzzle. She had had zero contact with health care: no hospitalization in the past fifteen years, no visits to friends in nursing homes, no family members working in medicine, no drug use. And her strain of MRSA—like the Harlem addicts, but unlike anything known in hospitals—was also resistant only to the methicillin family of drugs. Finally, in 1995, a physician in Jackson, Mississippi, wrote in a letter to a different medical journal that he had treated a six-year-old boy for a MRSA bone infection.[17] The boy lived a life in which he was surrounded by MRSA: his aunt worked in a local hospital as a clerk, and a second local hospital was in the midst of a MRSA outbreak. But the boy's strain of MRSA was not hospital MRSA. The MRSA strain in the Mississippi hospitals was resistant to many classes of drugs, just like MRSA strains in hospitals across the rest of the United States. But the strain of MRSA infecting the boy was resistant only to methicillin and its close relatives, just like the MRSA strains in the Harlem addicts, the Bronx grandmother, and the Chicago children who were starting to roll through Daum's hospital doors.

"The epidemiology of MRSA may be changing," the Mississippi doctor wrote. "Studies are urgently needed to define the true scope of this problem."

His plea was ignored. The medical establishment persisted in believing that MRSA was purely a hospital problem.

Tony struggled for weeks. Clarissa, camped out on the couch in his room, watched uneasily as the ICU nurses repeatedly checked his vital signs. Persistent fever indicates persistent infection, and his fever remained high. Once his belly ballooned hard and tight, and the staff feared his intestines might be dying. There was no time to get him to an operating room. The surgeons shooed out his mother, covered him with sterile drapes, and opened him up right there. "He was too unstable to move," said Dr. Robert Bielski, an orthopedic surgeon who began treating Tony within hours of his arrival. "He was really, really, really sick."

The belly problem turned out to be an accumulation of fluid, but it was another surgical insult to his slight, ravaged body. And there were more to come. Twice the orthopedic surgeons came back, peeled open the gaping wound in his thigh and washed it out again. In between, they plugged it with a "wound vac," a humming contraption that sucked blood and fluids out of the incision through a layer of sterile foam.

Tony had been taking vancomycin, one of the strongest drugs for MRSA, from the hour he arrived in the hospital. But the drug wasn't working. The continuing fever was one sign. Another was the constant presence in his blood of bacteria and other compounds that indicate inflammation. Day after day, they were there.

Vancomycin is an old drug, on the market for more than fifty years, and problematic in many ways. It has to be given intravenously. Its earliest formulations made recipients deaf, and its modern version can still be toxic to the kidneys at high doses. It kills bacteria slowly, and it does not penetrate equally well into all the body's tissues; it has an especially hard time killing bacteria in the lungs and bones. Vancomycin's patent protection expired in the 1980s, so no pharma company is likely to spend research money to improve it.[18] Still, none of the drugs that have come on the market to replace it work significantly better. It is an old, blunt tool, but the new ones in the toolbox are no sharper.[19]

Vancomycin is still effective against MRSA, despite decades of use. That is partly because it is a glycopeptide, a separate class of drug from penicillin and methicillin. It has a different chemical structure from the earlier drugs and attacks the bacterium in a different manner, and MRSA needed many years to evolve protections against those differences. But the drug's long record of reliability is also due to many years of sparing use. At first, that was because physicians had many alternative drugs to use. Later, it was because they had none. Vancomycin became a drug of last resort, doled out with extreme care for fear that widespread use would encourage resistance to develop.[20]

And resistance was developing. By the time Tony got sick, researchers worldwide were nervously tracking an uncommon strain of MRSA that they called VRSA because it was becoming so resistant to vancomycin.[21] There was a second bacterial opponent as well, a strain with an insidiously different adaptation that researchers called VISA, for vancomycin-intermediate S. aureus. VISA's adjustments to vancomycin's attack were subtle. Under the tests used in hospital labs, the bacteria appeared susceptible to normal doses, but hidden within the bacterial colonies were small groupings of staph that could resist that amount of drug.[22] That posed a double danger: When the usual dose was given, it would kill off the susceptible bacteria, but not the more resistant ones. And when the more-susceptible staph died, it left living space for the more-resistant staph to reproduce.

Tony had VISA. After seven straight days of vancomycin, his blood cultures still grew staph. Among the dozens of people working on him was Dr. Christopher Montgomery, an intensive care specialist who had done his residency at University of Chicago and stayed to join the faculty. He split his time between pediatric critical care and—as a break—the animal lab, trying to create a way to study MRSA in rats or mice. Tall and lanky, he was science-minded and precise, describing himself as "not good at feelings." Despite that reserve, Tony's lab results worried him. They seemed to say the MRSA strain infecting the boy would be vulnerable to the standard vancomycin dose, and yet the bug was manifestly not responding.

The surgeons had left Tony's leg open, but they had closed the incisions they used to tap his infected joints, because if it is left open to the air, a joint could dry out and freeze up. Montgomery thought that one of the infected joints might be harboring a more-resistant population of staph. He fretted about another possibility too. Tony was getting antibiotics through a central line, a catheter threaded into a vein below his collarbone. If the catheter had become contaminated with MRSA, it could conduct the bug straight to his heart. In the pre-antibiotic era, infections on the valves of the heart were almost uniformly fatal. So if Tony's strain was not responding to one of the strongest drugs, the outlook could be grim.

They took the line out—and for good measure, they switched him to a different new drug, linezolid, and added an old drug called rifampin as well. The choice of drugs suggested how few options there are for MRSA treatment. Linezolid, also called Zyvox, was the first of a new type of antibiotics called oxazolidinones that had only been approved by the U.S. Food and Drug Administration in 2000.[23] They were the first new class of antibiotics put on the market in thirty-two years. That newness had a downside: as a patent-protected drug, a course of linezolid could cost thousands of dollars.[24] And bacteria were already adapting to it. A year after linezolid's approval, researchers had already begun finding bacteria that were becoming resistant to it.[25] Rifampin, on the other hand, was a forty-year-old drug that could not be used on its own because staph develops resistance to it very quickly. It was one of many older drugs whose effectiveness against MRSA had never actually been studied. Physicians gave it, always in combination with other drugs, to gain whatever edge they could.[26]

Finally, after eight days, the team weaned Tony from the ventilator. Clarissa had been present the whole time, wrapped in the paper gown, gloves, and mask required for isolation precautions, napping on the couch or sitting by the head of the gurney. She was there when they woke him up. "I couldn't touch him, and at first he couldn't talk," she recalled. "But he saw me right away. I could see him trying to tell me to come and give him a hug."

Tony was still extremely sick. He was feverish and always sweating, but the worst problem was his left thigh. The surgeons had never closed the gash that reached from his hip to his knee because the long bone, the femur, wept infection into the muscles of his leg. Bone infections are usually concentrated in one section of a bone, but this one had spread throughout the shaft. The orthopedic surgery team kept bringing Tony back to the OR, where they washed out the wound with pulsing water jets. Then they drilled into the dead bone and ground it away from the inside. Bielski, the orthopedic surgeon, had never seen an osteomyelitis so bad. He tried not to think of what a weakened, reamed-out thigh bone would mean to a kid who loved sports.

The team changed the mix of drugs yet again to daptomycin and clindamycin, another new-old drug combination that they hoped might chase the last of the infection from Tony's blood. The switch precipitated a crisis: his fever soared, he broke out in a rash, and his blood pressure crashed. He was sedated and put back on the ventilator for a day, and then eased off it again while Clarissa watched anxiously. Over days and weeks, infections kept flaring, and it took twenty more surgeries to clean them out and repair the damage that staph toxins had strewn through his system.

The final surgery was a skin graft, to close the gaping wound in his thigh. On December 6, seventy days after arriving at Comer, Tony was finally well enough to be transferred out of the ICU.

* * *

After Daum's sentinel paper in 1998, researchers told themselves that his cases were an anomaly, some strange microbiological accident that was unlikely to happen anywhere else. But then physicians around the country began reporting clusters of illnesses that resembled the ones in the Chicago children. The new victims had not contracted their infections in a health care facility, and their bacteria had resistance patterns that were not like the health care strain.

An Army hospital in Honolulu reported cases in fourteen children and adults. There were fifty-nine sick prison inmates in Mississippi; fifty-three children in Corpus Christi, Texas; fifty-three other children living on a Native American reservation near the Canadian border; and several HIV-positive men in Los Angeles who were only identified because they were cared for by the same doctors.[27-31] The outbreak that got the most attention, though, was the smallest: four children in the upper Midwest who died of overwhelming infections between July 1997 and February 1999.[32]

The victims were a seven-year-old girl from Minneapolis, a sixteen-month-old girl from a North Dakota Indian reservation, a thirteen-year-old girl from a southern Minnesota town, and a year-old baby boy from North Dakota farm country. They had never met. None of them had been hospitalized before, and their families had no contact with health care facilities either. They had nothing in common but the MRSA that caused their deaths. It was resistant only to methicillin and the other beta-lactams. The fingerprints of two of the strains were identical, and the other two were very close matches—but all four were very unlike local hospital strains.

The children lived hundreds of miles apart, and investigators recognized their similarities almost by chance. Thanks to the cultural attitudes of several generations of Scandinavian immigrants, Minnesota has one of the best-funded public health systems in the country, with a crack microbiology laboratory, a medical complex that is the referral destination for the whole upper Midwest, and a corps of investigators who regularly lend a hand to less fortunate departments in surrounding states. Those consultations had created an informal network of lab scientists, physicians, and epidemiologists who shared stories back and forth about interesting cases. Some of them talked idly about the first case, who lingered in a Minneapolis hospital's ICU for five weeks with MRSA infections of the bones of both hips before she died. So when the second child was carried into a small North Dakota hospital with a 105-degree fever and died within two hours—and was discovered on autopsy to have MRSA abscesses in her brain, heart, liver, and kidneys—some of them remembered the first child as well.

Months passed, and then in January and February 1999, a third and fourth child died, both of them with lungs destroyed by abscesses and hemorrhage. By then a new member had joined the network. He was Dr. Timothy Naimi, an internist and pediatrician who had arrived in Minneapolis the previous August for a two-year posting with the CDC's Epidemic Intelligence Service, its frontline disease detectives. He heard about the first two cases from his new colleagues, and he was so struck by what they seemed to represent that he launched an investigation of his own, recruiting twenty-nine other physicians and scientists to help him. Almost exactly eighteen months after the Daum group's paper, they published details of the four children's cases in the CDC's weekly journal.

In the children's deaths, Naimi saw a warning—two warnings, in fact. The cause of their illness made it clear that the Daum group's cases in Chicago were not a one-time anomaly. And the fact of their deaths made it equally clear that the news of a new MRSA causing illnesses outside hospitals had not reached everyday physicians. All four of the children in Naimi's paper were taken to doctors by their families when they first fell ill. And all four had been given antibiotics—but because their doctors did not imagine that MRSA was out in the community, the children got beta-lactam antibiotics that would have had no effect on any MRSA strain.

"If you were a pediatrician or a family physician and a child came in with pneumonia or sepsis, you would not even have considered that MRSA could have been the source of infection," Naimi said. "We understood that we were seeing the very dramatic tip of a far larger iceberg."[33]

In Chicago, the iceberg was revealing itself much faster than elsewhere. There were so many cases that, when members of the Daum team spoke at medical meetings, listeners joked about "the Chicago disease." Even with the evidence from Minnesota and North Dakota, it was tempting to say that MRSA was simply different in Chicago. But it was much more likely that the University of Chicago researchers, knowing how the bug behaved, were better at identifying its unsuspected attack. About once every two months, they found the bug in children who were gravely ill with syndromes that had never occurred in hospital MRSA cases.

Among them were two unrelated infants, a one-month-old girl and a two-month-old boy, who were brought in one week apart in August 2000 with plunging blood pressure and lungs full of fluid, so septic the emergency room assumed they had meningococcal disease.[34] The boy recovered completely, but the girl was left with long-lasting neurological problems. There was a nine-month-old girl, sick at home for two days in April 2003 with what sounded like a cold, who died on her sixth day in the hospital despite being placed on a heart-lung machine. In 2004, there was a seventeen-month-old boy who arrived suffering from the same inflamed lungs and adrenal hemorrhage and died within twenty-four hours.[35] And in 2005 there was an eight-year-old boy, brought to the ER in shock after scraping his leg riding his bike, who spent more than two months on a ventilator, healing the five hundred holes that MRSA toxins had melted in his lungs.[36]

"With some of these kids, we have every supportive care at our disposal, all our resources, and we still can't do anything," Montgomery said. "We recognize it is staph very quickly. We treat with the appropriate antibiotics. And still we can't affect it." He paused. "I guess you could call that fear." What was especially troubling was that their numbers seemed to be increasing. In 2007, in the months that Tony was at Comer, four other children with severe MRSA disease came and went from the ICU.

Over years of study, the Chicago group began to understand why the new MRSA strain was behaving so differently. As their first PFGE results suggested, and later studies confirmed, the community-associated MRSA they found in their child patients was truly genetically different from the hospital variety.[37] It was a major research finding, but it only confirmed what they were already seeing in their patients: serious illnesses unlike anything that had been reported for hospital staph.

Tony was still the sickest they had seen. Even after he was discharged on December 9, 2007, he had to endure eight weeks in a rehab hospital on the shore of Lake Michigan. He had weeks of antibiotics still to take, and months of therapy. Finally, in March 2008, he put aside his crutches and for the first time in seven months put his full weight on his leg.

"Will the leg ever be completely normal? No," Bielski, the orthopedic surgeon, said. "Will he be able to play contact sports? Probably not. He has made a lot of progress, but I am sure we'll be following him for years to come."

Tony and his mother did not know, and likely would not have appreciated, that the many months of their ordeal had been just one tiny skirmish in a decades-long war between MRSA and the drugs that were developed to control it. Neither side gained the advantage for long. For every new pharmaceutical compound, the microbe developed a new defense, sometimes so quickly that the bacterium evolved resistance to a new drug even before the drug reached the market. Slowly, though, the bug was gaining the upper hand. Pharmaceutical research was running out of ideas and investment, while MRSA was finding new victims and causing illness in surprising new ways. It was a pattern that had been set almost fifty years earlier, with the first epidemic of drug-resistant staph.

CHAPTER 2

THE CLOUD BABIES

For weeks, Beverly Dieringer had been fretting over what she was about to say.

She worried about whether she could get the words out. She feared she would lose her nerve. She knew she would be judged. In 1962, even in liberal New York City, you could not say the words she was about to utter and not have people gasp and disapprove.

She had thought over the announcement she wanted to give, and now she thought it through again, rattling downtown on the No. 1 train, crosstown on the shuttle to Grand Central Station and uptown on the 6 train, and then walking across East Sixty-eighth Street toward New York Hospital with the wind from the East River lifting her hair off her face. She wondered about the reaction she would receive from the doctors and nurses who had been treating her family for two years for the boils they could not shake. Bev and her husband Ernie had scrubbed every hard surface in their apartment with antiseptic and bleached their sheets and towels until the fabric was fragile and crisp. The family had taken so many antibiotics that their daughters' teeth were dark and pockmarked. But as soon as one of them got better, someone would get it again: rashes, roaring fevers, painful, weeping sores. Ernie concealed them under long sleeves for fear he'd be fired from his job as a photo retoucher. Parents refused to let their kids play with twenty-two-month-old Eva and three-year-old Anna. Bev had been hospitalized for weeks while her sister-in-law Lorraine came out from Chicago to look after the girls.

And now there is this, she thought as she reached her doctors' office. It was too much. She was done.

"I'm pregnant," she blurted out. "If you can't make this better, I'm going to get an abortion. I'm not bringing another baby into all of this."

She expected them to shout, scold her, throw her out of the hospital. Abortion was a crime in New York State; it would not become legal there for eight more years, three years in advance of the rest of the country.[1] Even more than a crime, it was a scandal. Women obtained them all the time, of course, from sympathetic doctors if they were fortunate and from neighborhood amateurs if they were not. Sometimes the procedure was unremarkable; sometimes they survived with their fertility destroyed, and sometimes they died. But it was never easy, it was never discussed openly, and it was unthinkable that a woman would throw the challenge of an intended abortion in a physician's face.

She braced for their reaction, but they were courteous, and took her seriously. They would not discuss an abortion—it was illegal just to mention it—but they could offer a potential solution to the crisis that made her want one. They were working on a study, one of the doctors said, of families that were severely infected with staph. It was risky, and it was going to be difficult, but if the Dieringers were willing to try, the hospital might be able to alleviate their torment. They cautioned that there was no guarantee.

Bev walked slowly back to the subway, dizzy with relief and ebbing adrenaline. She could keep the baby. The girls would get better. Shakily, she picked her way carefully down the steps, wondering what the treatment would be like, and why she had never heard the word *staph* before.[2]

The family's ordeal started with Eva, the Dieringers' second daughter, though it took a while for anyone to realize she was the source. She was born at New York Hospital on September 1, 1960, and came home with Bev after the several days in the hospital that mothers were allowed back then. Approximately a week later, Eva developed a small spot on her butt inside her diaper—not diaper rash, but a single little pustule no larger than a pimple that rose and then healed within a few days. About a week after that, the family's symptoms began. Bev was breast-feeding Eva, and her right breast began to hurt. She looked and found a red spot, hot and somehow thicker-looking than the skin around it, but soft-edged as though it was surfacing from somewhere deep inside the breast tissue. Over a few days it expanded and flushed brighter, so warm that Ernie could feel the difference when he held his hand nearby.

Then the illness worsened. Bev felt feverish, and then confused and weak. Ernie came home from work and found her red-faced and staring, radiating heat like the coils of a toaster. He bundled her downstairs and flagged down a taxi, telling the driver to take her to her obstetrician. He followed her an hour later after begging a friend to watch the girls—but Bev was not at her doctor's office at New York Hospital. Five hours later, after scouring both sides of Manhattan, he found her lying on a bench in Columbia University's emergency room. She had told the taxi driver to take her to the hospital, but she had been too dazed to tell him which, and he had taken her to the closest one. Ernie carried her back to a taxi and got her crosstown, where the staff recognized the symptoms of a serious infection and sent her upstairs to be put in an isolation room. He went home, and found Anna, fifteen months old, hot to the touch too, her skin pebbled with an ugly flush. He headed right back to the hospital with her. They put her into isolation on a separate floor.

Bev and Anna were in the hospital for weeks. Every day, Ernie would ride uptown on his lunch hour to see them. Bev had a breast abscess that would not heal. The doctors had cut deep into her breast to empty it and put in drainage tubes, and every day the nurses would flush them out with a syringe that Ernie privately called the "turkey baster." To talk to her, he would step into a little closed foyer at the front of her hospital room, put on a gown, gloves, and a surgical mask, and slip in for a few minutes. Then he would take everything off, wash, go to Anna's room on the children's floor, and dress up again. He was terribly worried about Anna. She was scared, in the room by herself, and she wouldn't eat unless he sat with her when her meals came. She was not allowed to see her mother, but Ernie thought that was better; Bev's fever had risen so high that she had lost her hair and eyebrows.

He began to get sick himself: He ran a fever, and boils rose on his arms and back and buttocks. He kept his sleeves rolled down at work, hoping no one would notice how he shifted uncomfortably on his stool, and told no one. He worried doctors would keep him from visiting his wife and daughter, or isolate him too, and leave the family without any income and baby Eva—the only one of them not ill—without any parent at home.

After a few weeks, he got better. Anna did too, and the hospital let her come home. Bev's stay lasted six weeks; she came home mostly healed but scarred and tired and with her breast collapsed like an emptied balloon. Within a week, though, it was Eva's turn. A hot rash spread from her face and scalp across her trunk and back. Bev took her back to the hospital to be checked. A physician checked her over, ruled it a minor infection, and prescribed antibiotics to give her through a dropper.

And so it went, for two years. Ernie or Bev would feel the warning twinge of fever, or they would find one of the girls squirming restlessly and feel the heat rising off her skin. Every few weeks, they trekked back to New York Hospital, loading the toddlers into a taxi for the ride across town. The doctors would lance their boils and prescribe a different course of antibiotics. They took so many pills that Eva's teeth grew in shadowed and soft, pitted like something that had been dropped in acid. When she was two years old, her parents had them all pulled.

They kept bottles of a potent antibacterial soap at every sink and in the shower. They scoured the apartment, dry-cleaned the curtains, washed the slipcovers, bleached the sheets and towels. When the boils kept coming back, they packed up and moved to a new apartment. Even that made no difference. The boils kept coming. They moved a second time.

Until the day Bev announced she was pregnant, the Dieringers never heard the name of the organism that was bedeviling them. Perhaps that would not have mattered. There was not yet any patients' rights movement to encourage them to press for answers. They were not sophisticated and not likely to challenge their doctors; they only wanted relief. But it kept them from learning as well that they were not alone in their ordeal. They were part of the first epidemic of drug-resistant staph, an epidemic that had swept the world.

The antibiotic era had barely begun when its creators realized it contained the seeds of its own destruction.

Sir Alexander Fleming found the mold that made penicillin, the first antibiotic, growing by chance on a culture dish of staph in 1928. But he did not pursue the discovery for long, and a decade passed before the biochemists Howard Florey and Ernst Chain realized the drug's potential and began to develop it. In the summer of 1940, they gave it to mice they had experimentally infected with virulent staph, and saved them. In early 1941, they tried it for the first time in humans: first a housewife with terminal cancer, named Elva Akers, to see whether it was safe, and then a police constable, Albert Alexander, who was dying of an overwhelming bacterial infection that started with an insignificant scratch from a rosebush. The drug quelled the bacteremia, the overload of bacteria in his blood, and Alexander began to recover. But penicillin was so cumbersome to make that the researchers had been able to refine only a tiny quantity, and without enough doses to kill all the bacteria ravaging his body, Alexander relapsed and died. The next round of recipients, though, recovered from infections that would otherwise have killed them. That success persuaded U.S. pharmaceutical companies that they should manufacture the foreign drug. It was given to Allied troops starting in 1943, achieving battlefield cures of wound infections that killed soldiers in past wars. It was released to the public in 1944.[3]

Penicillin was a wonder drug, the first glimpse of the antibiotic miracle that would quell the ancient scourge of infectious disease, and its inventors were heroes. A portrait of Fleming appeared on the cover of *Time* in May 1944 over the caption: "His penicillin will save more lives than war can spend." The drug cut cases of syphilis and deaths from pneumonia by 75 percent and deaths from rheumatic fever by 90 percent. It did so well against serious diseases that its benefits were shared around for minor problems too. It was freely sold over the counter, in mouthwash, sore-throat lozenges, first-aid ointments, even cosmetics. In the United States, it was not made prescription-only until 1951.[4]

Fleming himself predicted what would happen next. In his speech accepting the Nobel Prize in December 1945, he said:

There is the danger that the ignorant man may easily under-dose himself and by exposing his microbes to non-lethal quantities of the drug, make them resistant. Here is a hypothetical illustration. Mr. X has a sore throat. He buys some penicillin and gives himself, not enough to kill the strepto-cocci but enough to educate them to resist penicillin. He then infects his wife. Mrs. X gets pneumonia and is treated with penicillin. As the strepto-cocci are now resistant to penicillin the treatment fails. Mrs. X dies.[5]

Fleming was sadly right, though amidst the joy over penicillin's impact, the research that would prove his prediction received little publicity. In December 1940, before the drug had ever been tested in a human, Chain and his Oxford University colleague Edward Abraham said in a letter to the journal *Nature* that the common gut bacteria *E. coli* seemed be evolving a defense against the new drug and was producing an enzyme that kept penicillin from working.[6] Two years later, Charles Rammelkamp and Thelma Maxon of Boston University demonstrated experimentally that staph bacteria could also develop protection against the effects of penicillin. They exposed successive generations of the bacterium to weak dilutions of the drug—the same situation in the laboratory that penicillin-laced soaps and face creams were creating in the outside world—and discovered that it took stronger and stronger doses to kill them.

Shortly afterward, they realized that penicillin resistance in staph was not only a laboratory phenomenon. Among fourteen patients who had received penicillin in experiments, they found four whose infections did not respond to standard doses. The staph strain in one, a forty-seven-year-old man with a deep bone infection, became sixty-four times more resistant to penicillin over eighteen days in which he received the drug.[7] And in 1944, researcher William Kirby of Stanford University in California discovered that staph's resistance to penicillin had not only moved across the country, but had spread beyond people who had been given the new antibiotic. He found an extremely resistant strain of staph in seven patients who had not received any penicillin at all.[8]

In late 1957, a group of very worried doctors gathered in Cleveland, summoned to an emergency session hosted by the American Medical Association. The object of their concern was persuasive evidence, from all over the world, that staph strains that had become newly resistant to penicillin were causing calamitous outbreaks in hospitals.

The first alarm rang in Australia in 1953. At the Royal Prince Alfred Hospital in Sydney, newborn infants who had left the hospital healthy, in the arms of healthy, beaming mothers, were being brought back, several weeks later seriously ill. If they were fortunate, the babies had only impetigo, scabby, crusted rashes of broken-open blisters and boils. The sickest were limp with fever, wheezing and blue with pneumonia, or comatose and in shock from the assault of bacteria coursing through their blood. Their mothers were ill as well, feverish and with hot, pus-filled abscesses swelling their already engorged breasts.[9] Penicillin was no longer the only antibiotic available. There were now several classes of antibiotics, from the penicillin derivatives to the sulfa drugs, the tetracyclines, erythromycin, and the antituberculosis drug streptomycin. But none of the new arsenal had a significant impact on the staph infections that made hospitals so concerned.[10]

The epidemic soon reached the United States. In March 1955, a young woman who had just given birth at a Seattle hospital returned to it several weeks later and died; she had succumbed to overwhelming sepsis that started with a breast abscess. Her death was not only tragic, it was shocking: the Seattle health department, which kept excellent records of local disease rates, had recorded no deaths from staph—in fact no serious staph illnesses at all—for years. Dr. Reimert Ravenholt, the department's chief of communicable diseases, began talking to local nurses. They all told him the same thing: babies and their mothers were leaving hospitals healthy and coming back ill. It took about eight days for the newborns to become sick, they said, and about fourteen days for the mothers. Tracking back, Ravenholt realized that the epidemic had begun sometime in the previous year. By mid-1955, he estimated, four thousand babies born in King County and 1,300 of their mothers had been made seriously ill by drug-resistant staph. Twenty-four babies had died.[11]

The outbreaks moved inexorably across the country. By 1957, they had a name. They were all caused by one staph strain, dubbed S. aureus phage type 80/81 after the results of a new laboratory typing method.[12] The pathogen was acquiring a fearsome reputation. It was much more infectious than was expected in staph. It almost always caused severe disease—not only in babies and their mothers, but in the elderly, people undergoing surgery, patients with severe flu, and anyone in a hospital who might be immunologically vulnerable or weak. And it was extremely adept at developing drug resistance. It caused so much havoc that, when the American Medical Association called for the Cleveland summit meeting, seventeen professional associations, medical schools, health departments, and federal agencies responded, including representatives of the Surgeon General, the Food and Drug Administration, and the predecessor of the Centers for Disease Control.[13]

The participants arrived in a state of concern bordering on panic. This was not how the antibiotic miracle was supposed to have played out. Infectious disease physicians were becoming accustomed to colleagues in other specialties jokingly hinting that they were dinosaurs, about to be made extinct by the new drugs' potency—yet here the drugs were already failing, and spawning a disease of a gravity none of them had seen before.

The reports they brought to the conference were grim. In California, 115 hospitals had fought outbreaks.[14] At Temple University Hospital in Philadelphia, 5 percent of the surgical patients developed ferocious wound infections.[15] The strain caused or contributed to 18 percent of the deaths at Iowa University's hospital.[16] "Nearly every hospital has experienced increased numbers of intrahospital-acquired staphylococcic infections that are antimicrobial-resistant, whether recognized or not," Dr. John Brown of the University of California, Berkeley, said during the conference.[17] New York researchers reported that they had surveyed hospital labs across the country for any data on local staph epidemics. In any hospital, they said, at least 78 percent of 80/81 strains were resistant to the three most important new antibiotics, penicillin, streptomycin, and tetracycline. In some hospitals, it rose to 98 percent.[18]

Reluctantly, the physicians admitted that their institutions might be in part to blame. After penicillin, pharmaceutical companies had achieved one new compound after another, all of them with dramatic effects on illnesses that had been intractable. Medicine had become so accustomed to their magic that it had started to relax the standards of rigorous cleanliness that once had been bedrock hospital behavior. Hospitals were using combs, brushes, thermometers, and the ink pads for newborns' footprints on multiple patients, and putting clean and dirty linen in the same transport cart. Hospital staff were sloppy about surgical masks, and instead of tying them tight to their faces would let them flap loosely from their necks. "A general breakdown has occurred in our 'aseptic conscience,' especially by the insidious development of short cuts, halfway measures and disregarded violations that physicians and nurses alike . . . would not have accepted twenty years ago," Dr. Mary Godfrey of the University of Iowa charged.[19]

The summit ended with a commitment to establish "infection committees" in hospitals, to monitor staff behavior and reinforce the necessity of sticking to strict aseptic techniques. It took a major hospital outbreak to demonstrate what could happen when newly resistant staph moved into a hospital that was careless and not clean.

Jefferson Davis Hospital, the cash-starved institution that took care of Houston's poor, was catastrophically overcrowded: its outpatient clinic, intended to handle two hundred patients a year, usually treated two thousand, and its maternity ward was built to take care of 250 babies a month but sent home 450. Segregation made it more crowded still. The eleven-story hospital had separate floors for whites and blacks, and while the white floors were often empty, the black floors were choked with patients. Moving black patients into unused white beds was inconceivable.[20]

Overcrowded, underfunded, pushed to cut corners and contemptuous of the minorities who made up almost two-thirds of its patients, "Old JD" was an outbreak waiting to happen. All it needed was the organism, and 80/81 supplied it. In March 1958, the hospital administration revealed that it had been struggling for months with an explosive epidemic. Sixteen newborns had died of drug-resistant staph, a revelation so out of scale even to the burgeoning national epidemic that the news made the *New York Times* and Surgeon General Leroy Burney called for an investigation.[21] Though the administration would not admit it for another year, the outbreak was in fact much larger: twenty-eight newborns and eight other children dead, and four hundred infected, all with staph 80/81.[22]

In Manhattan, the pediatric infectious diseases service of the New York Hospital was grappling with the same outbreaks as the rest of the country. Babies who left the hospital healthy were returning weeks later with abscesses, deep bone infections, and a severe staphylococcal pneumonia that was almost unheard-of in infants. The hospital was a teaching institution for Cornell Medical Center, and Dr. Heinz Eichenwald, an associate professor, noticed at once that the babies coming back to the hospital had been born at the same time and were falling ill at the same time. But after their births, the infants had been taken home to houses and apartments all over New York's five boroughs. The only thing they had in common were the New York Hospital nurseries where they had spent their first few days of life. It seemed clear that the nursery was where they had become infected.[23]

And that presented a problem that rigorous hospital cleanliness, the supposed answer to antibiotic overuse, could not prevent. Babies are born sterile, and within their first hours of life they acquire the first of the trillions of bacteria that live on and in humans throughout our lives. They pick up bacteria inside their mothers' vaginas, in their first swallows of breast milk, and from the air and surfaces in the delivery suites and nurseries. Some of those bacteria go on to form the stable populations called commensals that include S. aureus and other benign varieties of staph, and that most of the time coexist in harmony with their human hosts in a state called colonization. Everyday drug-sensitive staph was naturally present in the hospital environment and even in its air. But if virulently resistant staph 80/81 was there as well, then it was probably landing on the newborns too.

Eichenwald and his associate, infectious diseases fellow Dr. Henry Shinefield, figured out experimentally that staph 80/81 was a much more efficient colonizer than other strains of staph—in other words, it grew more rapidly, filling the microbiologically virgin territory of newborns' skin and mucous membranes so efficiently that more benign bacteria were crowded out. As few as five cells of staph 80/81 settling on a baby's fresh umbilical stump, or ten inhaled onto the lining of the nostrils, could grow into a stable population of the bacterium that would survive the antibacterial soaps used to bathe the babies daily.[24] The staph strain was such a potent colonizer that it persisted even if the babies were given a course of penicillin or erythromycin not long after birth. The antibiotics did not dislodge the bug; they only made it more resistant.

It was dismaying to discover that staph 80/81 established itself so quickly on the newborns if it reached them, because the researchers were also discovering how easily it reached them. They had proof that 80/81 was floating undetected in the air of their newborn nursery, and no doubt other hospitals' nurseries as well. The source was other babies, ones who gave no sign that they were ill.

It was already well-known that babies who developed staph skin infections shed the bacterium into the air on microscopic skin particles.[25] Hospitals controlled that threat by grouping babies with any signs of infection into small nurseries by themselves. If babies had no symptoms of disease, though, there was no reason to think that they were any risk to each other. That assumption turned out to be wrong: Eichenwald and Shinefield demonstrated that babies who were not obviously ill could shed staph 80/81 as well. Using an apparatus they developed called an impinger, a glass contraption that resembled a field hockey stick plunked into a vase, the two tested the air around apparently healthy babies in the New York Hospital nurseries, and discovered that half of the ones who showed no sign of staph 80/81 infection were colonized with the bug and shedding it into the air. The shedding could last for as long as five to seventeen days, which meant that the babies were not transmitting the resistant organism only to other babies. When they left the hospital, they brought it home, and infected their families as well.

Eichenwald and Shinefield needed to know whether anything triggered the shedding. They studied several of the 80/81 outbreaks among New York Hospital's newborn babies, plotting the cases against the time period in which they happened in what is called an epidemic curve. Three babies in particular seemed to have sparked outbreaks: they were colonized with 80/81, but had been transferred from one nursery to another before that was discovered. The researchers checked the babies, and in two of them found an intriguing clue: in addition to being colonized with staph, both babies had also had minor infections, caused by a cold virus in one baby and the virus for aseptic meningitis in the other. Since the infections were in their throats and staph colonizes the nose, the men hypothesized that there had been some synergy. Something about the viral infection must have changed the ecology of the nose and throat so that staph 80/81 could multiply faster, or could be sneezed out or exhaled in much larger numbers than usual. Whatever the mechanism, the viruses' presence was turning the babies into supershedders of staph 80/81—so efficient that, when they wrote the results up for a medical journal, the researchers called the newborns "cloud babies."[26] The journal that published the paper called the idea "preposterous"—no one had ever advanced an idea like it before—but also "an important, revolutionary concept."

It was 1960, well into the Cold War and its attendant anxiety over nuclear weapons and their inescapable deadly fallout—so the image of innocent newborns spreading clouds of invisible contagion seized the popular imagination. Eichenwald and Shinefield's conclusion was widely reported, and it tickled the creativity of cartoonist Charles Schulz, who memorialized the cloud babies in "Peanuts" on April 21, 1960. In the cartoon, Lucy reads to Charlie Brown from a press account: "Some newborn infants are highly infectious to others, and because they are literally surrounded by clouds of bacteria, they are called 'cloud babies.'" They both fall silent as Pig Pen, the child always portrayed in a swirl of dirt and dust, strolls into the frame. "Well," he says, "what are you looking at me for?"[27]

Despite the acclaim over their idea, the discovery that common viruses caused superspreading of resistant staph left Eichenwald and Shinefield with a conundrum. The point of their research was to find a way to prevent outbreaks of staph 80/81. Now they knew that any child with a cold had the potential to be a superspreader of the resistant bug, and that was a daunting finding, because in the long history of medical effort, no one had ever prevented the common cold. They decided their target would have to be the moment of colonization, when staph 80/81 made its initial contact with the newborns' skin and membranes. What they would need, they realized, was an even better colonizer: a staph strain that grew fast enough to prevent 80/81 from getting established, but was benign enough not to cause disease.

Their only problem was that no such strain was known to exist.

The inspiration for their search came from a medical mistake that had occurred fifty years earlier in Denmark. At the turn of the twentieth century, diphtheria—the "strangling angel"—was one of the most common and most feared diseases of childhood. It starts as a mild illness, with fever and a scratchy throat—but then it causes a tough, dark membrane to grow across the throat, and chokes to death one-fifth of children who contract it. Epidemics were so dreaded that patients were locked in "diphtheria wards" until they recovered. In 1909, a physician named Schiotz made a bad diagnosis, and put a man who was developing a sore throat into an isolation ward. The man did not have diphtheria, though; he had a staph infection of the throat. And he never contracted diphtheria; the staph prevented the second infection from taking hold.[28] Schiotz was so impressed with the accidental discovery that he tried spraying live staph bacteria into the throats of patients who had recovered from diphtheria but had become long-term carriers. They were considered such a risk to the community that they had been forcibly isolated in the local hospital: one man had been there for three months, and a woman had been confined for two. Sure enough, the staph bacteria in the solution dislodged the preexisting diphtheria. Schiotz was able to free at least six people—some reports say forty—from their carrier state.[29]

The innovation passed out of use once diphtheria vaccine was developed in the 1920s, but it remained in medical lore. Eichenwald and Shinefield wondered whether they could use the phenomenon, dubbed "bacterial interference," to dislodge 80/81 from the babies in the New York Hospital's nurseries—if they could find a safe, fast-growing strain.

Chance handed them what they needed. The two researchers had begun swabbing every baby born at the hospital before they went home to check whether they were colonized with 80/81. The swab results helped the men predict whether the babies would become ill—48 out of every 100 colonized with 80/81 did—and whether they posed a risk to their families. The hospital had a number of small nursery rooms, and as the surveillance went on, the two noticed something odd: almost none of the babies who spent their first week of life in one particular sixteen-bed nursery were colonized with 80/81.

They swabbed all the staff who worked in that nursery, and in the nose of one nurse, they found a staph strain that was new to them. They dubbed it 502a. They rechecked the babies from her nursery and found they all shared that same strain and only that one. It had colonized the babies first, and kept 80/81 out. But without molecular analysis—still decades in the future—there was no way for them to tell whether this 502a strain that seemed just as infectious as 80/81 would be just as virulent as well. Their only option was to send the babies home, and wait nervously to see whether they came back.

Almost none of them did. A few returned with superficial infections, small bubbles in the skin around their healing belly buttons, the spot where 502a probably had begun growing. But there were no boils, no sepsis, no devastating pneumonias—and just as important, there was no illness in the mothers either, or in the rest of the babies' families. Fate, or blind luck, had given Eichenwald and Shinefield the weapon they needed against 80/81.

To check the new strain's potency, they challenged it directly. They took babies who had spent their first sixteen hours of life in the room where 502a was circulating and moved them to one where more than half of the babies routinely contracted 80/81. Of the fourteen babies they transferred, twelve never acquired the disease-causing strain. That gave them enough data to try giving 502a to newborns deliberately. Within the first hour after a baby was born, they dripped a tiny amount of 502a—100 cells at most—into the nose and onto the fresh umbilical stump. Nine out of every ten became colonized with 502a and acquired its protection, and did not contract 80/81. Even better, when the babies went home, they passed the protective bug to their families as well.[30]

Results that striking were difficult to keep quiet, and other hospitals began asking the Cornell team to help them eradicate their own outbreaks. Shinefield, with no faculty position, was freer to move around the country. He brought samples of 502a to Cincinnati, Atlanta, New Orleans, and elsewhere—inoculating more than four thousand children and short-circuiting epidemics in eight cities.[31-35] Other hospitals learned about the technique in conference presentations and tried it on their own. The conferences could be raucous; other pediatricians criticized the Cornell team's ethics and challenged their statistics.

But parents of babies born at New York Hospital after the 502a program began—almost four hundred babies over several years—were thrilled with its success, grateful to be spared what had become a notorious ordeal. Then families whose children were born earlier, and who had struggled for years with the virulent staph strain, began begging the hospital to do something for them as well. The researchers had not envisioned colonizing families, because it was much more complicated. They would have to start, not by filling the blank slate of a newborn's naïve immune system, but by erasing the long-lived bacterial population of an adult's nose and skin. There was no guarantee it would work. On the other hand, the families were pleading. Some had abandoned their houses and livelihoods to escape the infection, unsuccessfully. One couple had divorced. Shinefield did some preliminary research, and thought it might be possible.[36]

And at about that time, Bev Dieringer walked into New York Hospital to defy the team of doctors who had been treating her family for staph for two years. The head of the team was Heinz Eichenwald.

In the heavy accent of his Swiss childhood, undiluted though he had arrived in the United States at age eight, Eichenwald explained what the Dieringers would have to do. They would all of them—Bev, Ernie, Anna, and especially Eva—have to report to the hospital every day for ten days to take heavy doses of antibiotics. They would get one day off, to let the drugs wash out of their systems; on that day, they would have to sequester themselves at home to avoid any accidental infection. Then, on day twelve, they would come to the hospital for a dose of 502a staph, dabbed into their noses as the researchers had done for the newborns. If the dose took, it ought to replace the virulent staph that had colonized them. If the replacement was complete, they ought to be cured.

Neither Ernie nor Bev questioned the plan for a moment. It was a relief just to have an explanation for their ordeal, and a name for it. To be offered an escape was overwhelming. *Here's a straw*, Bev thought. *I'm grabbing for it.*

They went to a meeting in a hospital conference room that was attended by other families as well. Ernie counted thirty-seven parents and toddlers and older children. He wished he had met them before, and then realized ruefully that they could never have been friends. For two years, his own family had been rootless and furtive. The girls did not have playmates over, and the parents did not have dinner guests; even friends who were nurses, who had lived in their first apartment building, declined to enter their doors. The other families would have been in the same position, awkwardly isolating themselves in the busiest city on Earth. Now, perhaps, they could all have normal lives again.

The moment of remedy was almost anticlimactic: a Q-tip dunked in a solution, a brief twirl in each nostril, and they were done. The hospital would be back in touch.

And they *were* done. The treatment took. None of the Dieringers ever again experienced a single boil. They had other health problems, Eva in particular, possibly because of the many antibiotics the family took over the two years of turmoil. She had neurological difficulties, pneumonia, conjuctivitis—and she bruised so easily that physicians once interrogated Bev to make sure no abuse was taking place. She was allergic to almost every antibiotic on the market. Anna was unaffected. So was

Joan, the baby that Bev was carrying when she challenged Eichenwald to heal her family. Joan was born healthy in December 1962.

The New York Hospital team checked on the family regularly: every few weeks, then every few months, then once a year for more than a decade. Free to restart their lives, the Dieringers moved to Connecticut. They put their staph ordeal out of their minds, and never told the story to their daughters, who were too young when it happened to remember it as they grew up. Eva only learned of the family epidemic when she sought her childhood medical records in 2007, trying to understand her lifelong health problems. She was stunned at her role in the hidden history and was determined to find out more. She tracked down Eichenwald, who left New York in the mid-1960s and had a distinguished career at the University of Texas Southwestern Medical Center in Dallas. He had just retired. Eva wangled his number from his former secretary, and called him.

"I want to thank you for saving my family," she said.

Shinefield moved away as well, to Kaiser Permanente in California, where he founded the medical giant's Vaccine Study Center and kept up his research into bacterial interference. As years passed, 502a came to seem less miraculous. It caused several outbreaks of mild skin disease in newly colonized infants, and was blamed for both a vicious case of recurrent boils in a Tennessee girl with chronic skin disease, and the death from septicemia of a premature infant in Texas.[37-40]

Chiefly, it came to seem like a curiosity—because even as the Dieringers were suffering through their two-year outbreak, a new compound was making it onto the market that would vanquish staph 80/81 for good. It was called celbenin and made by Beecham Research Laboratories in England. The manufacturers announced its advent in a back-page ad in the British *Quarterly Journal of Medicine*. "Effective against all resistant staphylococci," the ad promised. "Resistance unlikely to develop."[41]

The drug made good on its promises, at least at first: it controlled staph strains that were resistant to the original penicillins. Physicians could tell it was working, because 80/81 simply disappeared. At the New York Hospital, where Eichenwald and Shinefield were still doing painstaking surveillance to isolate every staph strain, it vanished from the swabs. Other hospitals saw the same. The National Institutes of Health, just outside Washington, D.C., ran its own hospital entirely devoted to research into odd, orphan, and life-threatening diseases. It was called the Clinical Center, though everyone in American medicine just called it "Building 10." Because every patient in Building 10 was there for some dire reason, NIH did meticulous microbiology to track any pathogen that made it through the doors. And 80/81 was one of them, but only for a while. It initially appeared at NIH in 1959, four years after it killed that first young woman in Seattle, and was completely gone by 1964.[42]

Celbenin was the first synthetic penicillin derivative. It had taken three decades from the first isolation of natural penicillin for pharmacology to understand how the compound had disarmed staph's defenses, and how staph had overcome that mechanism in turn. The secret of the drug's success was that it killed bacteria by interfering in the synthesis of new cell membranes when bacteria reproduce, weakening the walls so that the bacteria could not maintain their internal pressure, but burst apart and died. Staph's response to that mechanism was an enzyme—the same one that Chain and Abraham had spotted in E. coli in 1940—that rendered penicillin inactive by cleaving through a feature of the drug's chemical structure, an arrangement of atoms called the "beta-lactam ring," before it could bind to the cell.

Celbenin was an engineered attempt to foil that response. Pharmaceutical chemists had started with the original penicillin molecule and then added to it, adorning it with bulky chains of compounds intended to protect the beta-lactam ring from the enzyme.[43] Because the drug was man-made and new, they were confident that resistance, if it occurred at all, would be many years down the road. The compound was renamed methicillin, and released for sale in 1959 and 1960 in the United Kingdom and the United States.

But in January 1961, a British bacteriologist named Patricia Jevons sent a letter to the *British Medical Journal*, informing them that she had found resistance to methicillin in three bacterial samples sent to the United Kingdom's National Staphylococcus Reference Laboratory. The samples all came from one hospital: from a man who had had a kidney removed, a nurse who cared for him and developed a skin infection, and another patient who was treated in the same ward two weeks after the man had gone home. It took doses of methicillin five times larger than usual to quell their infections.[44,45] The era of MRSA had begun.

34

CHAPTER 3

THE SILENT ATTACK

"And just remember—Be Big!"

"Big Jay" Sorensen, morning drive-time radio host, clicked off his microphone, slid off his headphones, and winced. His stomach hurt.

It was a sizable stomach, to be frank. Big Jay had earned both his nickname and the motto that served as his daily sign-off. He was larger than life in personality and in shirt size, and he had the booming voice and gregarious nature to match. He was fifty-two years old and he had been working in radio since he was seventeen, knuckling his way up from news stringer to board operator to program manager to what he liked to call "classic-hits musicologist." He had worked at WNBC-AM, a legendary New York City station that birthed the biggest names in 1970s and '80s radio: Wolfman Jack, Howard Stern, Joey Reynolds, Don Imus. He developed a following of his own, people who recognized the difference between the Dave Clark Five and the Five Man Electrical Band and knew that the first place in the United States where one of the Beatles played in public was an Illinois VFW Hall, not the Ed Sullivan Show. Now he was back in Manahawkin, New Jersey, thirty miles from the oceanfront town where he grew up and got his first job, playing early-rock classics and bantering with Anita Bonita, co-host of "Big Jay and Anita in the Morning" on WJRZ-FM.

To get to the station in time to run through newspapers and websites—fodder for the interviews and calls that were an essential part of the show—Sorensen got up at 2:45 every morning. So when he finished a shift, he expected to be tired and mentally wrung out. But he'd never felt like this before. There was a knot of pain just below his navel, so intense that it was hard to stand up from the big chair he sat in to broadcast. He unfolded himself carefully, straightening as much as he could, and made his way slowly out of the studio and into his car for the midday drive home. It was late August 2005, and his wife Andrea, who worked part-time at a tax-preparation office, was waiting for him for lunch. They never got lunch. "We have to go to the hospital," he said. "I don't know what, but something's terribly wrong."

The local emergency room identified the cause pretty quickly: Jay had diverticulitis, small, balloon-like sacs protruding from the surface of his large intestine into his abdominal cavity. One of them was swollen and infected, and on ultrasound it looked like it had ruptured. That was what was causing the piercing pain. The staff admitted Jay to the hospital, and told him and Andrea that they would try him first on antibiotics and pain medication, to see whether the infection would clear and shrink back on its own.

It did not. After a week in the hospital, his own doctor told Jay to prepare for surgery, a bowel resection that would take out two feet of his colon and stitch the snipped ends back together to make his bowel function normally again. Diverticulitis was a common problem, the doctor said reassuringly, and bowel resection was a routine surgery. He would be in the hospital for about a week afterward and then go home. Jay was a healthy man, for all his size, and his doctors predicted his recovery would go smoothly. He would be back to normal in just a few months.

In fact, it would take years.[1]

The surgery took place on a Monday, and by Tuesday Jay was up and hobbling through the hospital hallways, shuffling along in rubber-bottomed booties and hanging on to the IV pole for support. He hoped to go home on the weekend, but on Thursday he started to feel sick, and on Friday, his colorectal surgeon came by, spotted a bubble of fluid behind his stitches, and drained it. By Saturday, Andrea was nervous; Jay kept asking for more pain relief, but once he downed the meds he babbled and made no sense. On Sunday, he seemed better, though, and when Andrea mentioned that their daughter Tracy wanted to bring the grandchildren to visit, Jay levered himself to his feet and hunched over the sink to wash his face and shave.

"He bent over, and all this orangey-brown stuff came pouring out of his stomach, out of the wound," Andrea said. "I made him sit down and I screamed for the doctor. The one who was covering for the weekend came in and looked and said, 'He's got a site infection. Your doctor can look at it tomorrow, it will be fine.'"

The next day, his doctor paged the infectious disease physician on call, and she arrived quickly and prodded the wound. She ordered some cultures and, knowing they would take twenty-four hours or more, started Jay preemptively on vancomycin, the MRSA drug of last resort.

It was a smart assumption. When the culture results came back the next day, they verified a MRSA infection. Even worse, they indicated a second, equally serious drug-resistant bug: vancomycin-resistant *Enterococcus faecium*, known as VRE. As its name suggests, E. faecium is a gut bacterium; for about ten years, it had been causing ferocious bouts of hospital infections. VRE no longer responded to vancomycin and could be treated only with older, more toxic drugs or with newer, more expensive ones that not all patients could tolerate.[2]

Jay had just become a victim of both hospital-acquired infection and collateral damage in the battle to contain it. VRE had become vancomycin-resistant in the same years that MRSA had begun to intensify, in hospitals and also in the serious community infections that Robert Daum and others had uncovered. More potent strains of MRSA had led to greater use of vancomycin. That suppressed MRSA to some extent, but encouraged the development of VRE instead.[3] The result was that Jay needed to be loaded immediately with two potent antibiotic regimens, one to combat MRSA, the other to beat back the VRE that the MRSA drugs could not contain.

For two weeks, he stayed in the hospital, watching nurses and technicians shuttle in and out to take his blood for cultures and hang fresh IV bags on the beeping infusion pumps by his head. There was little else to watch, and almost no one to talk to—because to keep the infection from spreading, the hospital had put Jay in isolation. That meant every staff member and visitor had to wash and put on gown, gloves, mask, and shoe covers before entering his room and discard them and wash again after leaving. The routine was so onerous that it discouraged casual visitors and reduced traffic to only the staff who most needed to come in. Andrea was there all the time, day after fifteen-hour day, wrapped in a paper gown over her street clothes, fidgeting with the mask elastic that hooked uncomfortably over her ears. Fearing infection, she forbade friends and family to visit. Grandchildren were out of the question.

On the day before Jay was to be released, she got a call at home. It was early Sunday morning, and she was getting ready for church before heading to the hospital for another long vigil. Jay's surgeon was on the line with bad news: the bugs were not being beaten.

"He said, 'I hate to do it, but we have to go in and clean him out,'"

Andrea said. "So they took him in the next morning and opened up the incision, and basically power-washed his insides."

When the surgical team opened the two-week-old incision, they found that the MRSA infection had gone further than they thought. It was not only immediately beneath the vertical slice through Jay's abdomen; it was also in the stitched-up resection of his colon, on the lining of his abdominal cavity, and on the outer surfaces of his organs. The high-pressure lavage they used dislodged the infection, but it negated most of the healing his body had managed in the past fortnight. When they closed him up again, they realized he was facing several more weeks in the hospital, weeks that he would be spending in isolation until the last-chance antibiotics brought the infection under control.

Jay finally made it home on a Friday, six weeks after he had gone in for surgery. Two nights later, he woke Andrea, worried about a sharp pain in his back. He was having trouble breathing too, but he dreaded going back to the hospital. It took the combined arguments of Andrea and the home-health nurse who visited the next day to get him there. He relented at last, and Andrea drove him, fearing he had pneumonia after so many weeks on his back. But the news was worse than that. He had a pulmonary embolism, a blood clot that starts in a major vein, usually deep in a leg, and travels through the circulatory system toward the heart.

Pulmonary embolisms are a life-threatening emergency and require immediate treatment with clot-dissolving drugs that are given by IV. Jay was admitted, and when his ER bloodwork was returned, was put back into isolation as well, because the cultures showed MRSA was still present in his blood. It took another three weeks before he could leave again.

He had been home several weeks and was dressing carefully to avoid jarring the still-tender incision when Jay noticed a lump in his groin. He went back for a check-up, and his surgeon confirmed it: the bulge was a hernia. The two surgeries through his abdominal wall had weakened the muscles, and his internal organs were protruding through them. Jay was worried by the diagnosis, but more worried about risking surgery again; through the winter of 2005 and the spring of 2006, he nervously waited as the lump got bigger and more painful. By summer, his colorectal surgeon felt it could go no longer. He recruited a plastic and reconstructive surgeon, Mike Nagy, who repaired the hole with a patch of preserved cadaver skin.

It was a solid repair, though it left Jay with a set of scars shaped like an anchor—the original vertical scar from the previous summer, and the smile-shaped new scar that went from hip to hip. But while the operation was a success, the recovery was not. The MRSA recurred.

"They'd done some blood work, and he was sleeping," Andrea said. "I saw the nurse come in with an IV bag I recognized, the bag of vancomycin, and I thought, *Oh no.*"

Jay went home after a week, still on IV antibiotics—but the combination of MRSA and the blood thinners he took to prevent more emboli kept the long incision in his abdomen from healing. In September, Nagy had him come back for more surgery, his fourth, to remove dead tissue around the sides of the wound. That left Jay with a hole in his abdomen too wide to close surgically—"You could fit a melon in it," he said, "a cantaloupe"—and so Nagy ordered home care. That meant a "wound vac," a contraption that had to be set up twice a day by a nurse. A thick layer of foam was trimmed to fit the wound and pushed gently inside the hole, topped by a sticky film to seal it. Then a tube with an adhesive collar was pushed into the foam and connected to a portable vacuum the size of a toolbox. When it was turned on, the vacuum hummed and slurped, sucking away wound fluids to encourage new tissue to grow. It made Jay grateful the nerves in his abdomen were dead.

For three months, he carted the box around with him, tolerating the gurgling. "It's my purse," he joked. "My purse that makes noise." He went back to work at the radio station, leaning back as far as possible in the most flexible chair because the wound made it too hard to sit upright, and levering the mike close to his mouth so it would not broadcast the slurping. The machine was a disgusting companion, but it was effective: slowly, the wound was closing. But Jay was still on high doses of antibiotics, and eventually they took their own toll. He developed thrush, a creamy white coat of fungus over his mouth and throat that for weeks made it too painful to swallow anything but ice chips and Gatorade. On a day in January 2007, stopping at his daughter's house on his way home from work, he collapsed. She took him to the ER. Andrea rushed to meet them there and bumped into a friend who is an emergency nurse.

"He's going to be here a while," the friend told her. "Does he have a spare battery for the wound vac? Why don't you go home and get it?"

It was months before her friend confessed what that really meant: she thought Jay was about to have a heart attack, and she didn't want Andrea to see.

* * *

From its earliest emergence, MRSA was a hospital phenomenon, though it took a while for medicine to admit the bug had settled there. The three initial cases found in 1961 by the British national program of microbiological surveillance were assumed to be an anomaly. British scientists put the episode down to random genetic accident, the slot machine of staph's relentless evolution happening to line up in a particular way at a particular time. They were wrong, though. The cases were not a quirk. They were a warning of how quickly methicillin resistance had emerged and was spreading in the country where the drug was used first.[4]

Eighteen months later, in September 1962, doctors at the Central Middlesex Hospital in northwest London gave emergency treatment to a twenty-eight-year-old railway worker. In a terrible accident, he fell on the tracks and was run over by a train that cut off his left leg below the knee, crushed his right shoulder and ribs, and ruptured his spleen. He had been in the hospital for three days when he developed a searing 106-degree fever. The staff rushed to give him antibiotics, and started him on penicillin, streptomycin, and methicillin. Five days later, realizing the infection was subsiding more slowly than it should, the hospital cultured his blood for bacteria and found a staph strain that they judged "weakly resistant" to methicillin. But by only two days later, the same strain had grown significantly more resistant. They dropped the methicillin and switched him to two other drugs, but he had to stay on them for two more weeks before the infection was finally quelled.

There had been no methicillin-resistant staph at Central Middlesex up to that point, and the hospital was bewildered where it might have come from. The staff checked lab results from all the other patients, and found that the same "weakly resistant" strain had been recorded on September 4 in a woman recovering from a surgical hip fusion. That was the same day the railway worker arrived at the hospital. The hospital reluctantly concluded that the staph strain had not arrived with the gravely injured trauma case. The institution had given it to him.

"The appearance within four days of two strains of Staph with identical antibiograms . . . and both resistant to methicillin is unlikely to be coincidental," the hospital pathologist gravely noted in a report. "It is therefore accepted that cross-infection occurred."[5]

Someone must have conveyed the methicillin-resistant staph from one patient to the other, because between the hip fusion and the amputation, neither the woman nor the railway worker could walk. The hospital said that it was not possible to identify who might be responsible, even though the day of the woman's surgery was an end-of-summer holiday when the institution was running on a skeleton crew. The few staff who worked that long weekend were given days off later in the month, and they were still absent when the two staph infections were diagnosed. When they returned to work, they were checked, but no staph was found.

What the hospital was looking for, in checking the heath care workers, was staph colonization, the asymptomatic carriage that was the key to Eichenwald and Shinefield's treatment for penicillin-resistant staph 80/81. It had been known since the 1930s that people could carry staph without being made sick by it, usually on the mucous membranes just inside the nostrils.[6] Staph can also survive on the moist skin of the groin and armpits, and on dry exposed skin as well if the skin's immune protection has been disrupted by eczema or similar diseases.[7] Up to 30 percent of adults carry staph persistently, and many more can be transiently colonized for days or weeks if they are exposed to the bug.[8]

Despite the revelation of the cloud babies, it was not yet understood how often colonization led to disease. People who developed staph boils and sties tended to be infected with the same strain they were carrying in their noses—suggesting they had infected themselves—and also were more vulnerable to wound infections after surgery.[9] And carriers could infect others even if they were experiencing no symptoms themselves. In 1958, the brand-new U.S. Centers for Disease Control determined that the source of thirteen staph infections at St. Joseph's Hospital in Atlanta—several of them life-threatening, one of them so serious that the patient lingered in the hospital an additional five months—was a surgeon who was an unwitting nasal carrier.[10]

Staph colonization is remarkable. There is no other disease-causing bacterium that is carried, without causing infection, by such a substantial slice of the population. The arrival of antibiotics altered that stable balance. They put the bacteria under what scientists call "selective pressure," creating a microbiological battlefield on which the ones that developed resistance would be more likely to survive. Researchers soon noticed that the hospital patients who were colonized with drug-sensitive staph became carriers of drug-resistant staph instead once they were given penicillin.[11] Then they found that hospital workers, working in an environment where both pathogens and antibiotics were common, were even more likely to carry drug-resistant staph than their patients were.[12] It was too soon to know whether new methicillin-resistant staph strains might colonize patients more quickly, or persist in their nostrils more durably, or move from one person to another with more facility than penicillin-resistant strains had. But it was evident that broad use of antibiotics not only caused drug-resistant infections; it also made people who had no symptoms of infection into silent carriers of drug-resistant strains. The first hospital outbreak of MRSA demonstrated how significant that would be.[13]

Queen Mary's Hospital for Children, the first dedicated pediatrics hospital in England, boasted an odd but beautiful layout, the relic of an Edwardian-era hospital-building boom. It comprised three ranks of twenty-four single-story buildings, laid out on 136 acres of lavish park-like grounds, and housed 850 children undergoing medical care, tuberculosis quarantine, and long-term rehabilitation. It also had a busy and curious microbiology lab that began scouting for methicillin resistance as soon as the hospital began to use the drug.

In March 1961, the hospital found its first methicillin-resistant isolate in a baby who had been admitted in January with a drug-sensitive staph infection. She got penicillin, and the infection resolved, but she remained in the hospital for unrelated reasons—and when she was checked two months later in a routine survey, she had become a carrier of methicillin-resistant staph. The next month, the same strain had spread to four other children. Two of them were only colonized, but two were infected, one within a patch of eczema and the other with an ear infection. The hospital tried to cordon off the strain, isolating the children with infections, closing the building where they lived to all visitors, bathing the patients and staff with antiseptics and inserting antibiotic ointments into all of their noses. It was too late, though: by May 1962, the bug spread through the hospital, to eight of its forty-eight wards, infecting thirty-seven children and one nurse. One child died.

The hospital had mounted what it called, in a research report, a "fairly vigorous prophylactic campaign" to contain the strain. Yet the outbreak persisted for fourteen months and spread with ease between widely separated buildings on a campus where few patients were well enough to walk that far. That raised the possibility once again that an unwitting carrier—a patient, a parent, or a health care worker—could be the vehicle for transporting a resistant strain.

* * *

Throughout the 1960s, British and European hospitals reported increasing amounts of MRSA, and so did Australia. Sydney's Royal Prince Alfred Hospital built a new surgical ward expressly to reduce outbreaks.[14] But by January 1967, there had been no MRSA outbreaks in any U.S. hospital. Doctors at Boston City Hospital, the overcrowded charity institution in the city's South End, even crowed in a medical journal that hospital infections caused by staph had "decreased markedly."[15]

Just weeks after publishing that account, Boston City discovered the first MRSA cases in the United States' first outbreak, which would spread through the hospital for almost a year. It started with patients in a single, small medical ward. One arrived coughing from a lung infection, and two days later the hospital found the same strain in another patient in a nearby bed. Ten days after that, a third patient in the same ward was found to be infected, and a fourth patient in a separate ward. MRSA percolated through the building: one case in March, two in May, five in June, one each in August, October, and November, two in December, and three in January 1968. The outbreak comprised two waves of infections—bronchitis, pneumonia, urinary tract infections—that each seemed to be triggered by elderly patients who had arrived sick from nursing homes. But the bacterium also seemed to be moving back and forth between patients and staff. The eighteenth case in the outbreak was a nurse who was only colonized, with no symptoms, who had treated the previous four cases. The patients were all men, housed in one small room with four beds. Two had pneumonia and were coughing, and one had a tracheostomy that needed regular suction. It was impossible to determine who was infecting whom. Tracing the bug's path through the building, the hospital's investigative team issued a report that sounded like a prophecy: "It seems reasonable to predict that methicillin-resistant staphylococci will become more widespread in the United States, and may present some clinical and epidemiological problems in the future."[16]

They were right. MRSA vanished from sight for several years, but in the mid-1970s it began to hopscotch the country. In 1976, twenty-three patients at Hartford Hospital in Connecticut, almost all of them recovering from surgery, became infected in less than a year and developed very serious illnesses: pneumonia, osteomyelitis, lung abscesses, bacteremia. Three died.[17] In 1978, sixty-one patients at Memorial Hermann Hospital in Houston contracted MRSA, most of them surgery patients or burn victims.[18] That same year, the University of Virginia's hospital admitted a patient with a chronic skin condition and a heart-valve infection who brought MRSA with him. The bug leapfrogged through the hospital, from that patient to a second patient via their shared physician; from the second patient to a third, fourth, and fifth by the close proximity of their beds; and from the fifth patient to five more when he was transferred into intensive care. Within six months, sixteen patients were infected, fourteen more became carriers, and three died.[19]

By 1980, 109 hospitals that responded to a national survey had treated at least one case of MRSA. Dismayingly, many of those cases had blown up into outbreaks. Thirteen institutions said they had had fewer than ten cases; but thirty-four said their outbreaks had involved fifty cases or more. Only thirteen could say they were confident that they had chased the bug completely from their buildings, and twelve of those were hospitals where the outbreaks had been very small.[20]

A picture was starting to emerge of who was most vulnerable to MRSA. They were patients who were very young or elderly, or severely or chronically ill, all conditions in which the immune system is not robust. In one outbreak at St. Louis University School of Medicine, 70 percent of the fragile newborns in the neonatal intensive care unit were colonized.[21] Patients were in greater danger if the protective barrier provided by their skin had been breached by surgery or burns; at the City of Memphis Hospital in Tennessee, burn victims were infected by each other and by the hydrotherapy machinery used to debride their wounds.[22] And they were more likely to be colonized or become infected if they had received any antibiotics, because any sensitive bacteria they carried were killed by the drugs and replaced by the hardier resistant organisms.[23,24]

It was also becoming clear how MRSA reached vulnerable patients in order to infect them. The organism colonized patients, but it also survived on hospital surfaces—not just hard surfaces such as counters and bed rails, but bed curtains and doctors' coats as well.[25] And as Queen Mary's Hospital and Boston City Hospital had suspected, it colonized health care workers, sometimes transiently, sometimes long-term. An outbreak among sixty-six patients at two Oregon hospitals was traced to a single colonized physician who worked at both of them.[26]

There was suddenly so much MRSA, in so many different hospitals, that patients' own colonization could no longer be blamed for it. Something was transporting the bug from its many hiding places—patients, staff, the hospital enviornment—to the many vulnerable patients it attacked. With outbreaks increasing across the country, hospitals reluctantly faced a difficult realization: sometimes the vehicle was their own health care workers. Even when there were MRSA outbreaks in their own institutions, health care workers were not washing their hands.[27]

All of those factors—rising incidence, vulnerable patients, unrecognized colonization, inattentive care—came together in one terrible outbreak that began in 1980. It forever established MRSA as a deadly hospital threat.

In June 1980, a sixty-three-year-old man from Washington State who had been badly burned in Texas and was treated at a Houston hospital that had been combating MRSA in its burn unit, was transferred to Harborview Medical Center, the public hospital in Seattle. Harborview knew he had MRSA; it grew from a wound culture taken when he was admitted, and from his sputum and blood in the days that followed. But MRSA was not new in Seattle; doctors there had experience treating it, and the hospital must have believed it knew what to do.

The man died twelve days after arriving. In those twelve days, he set off a chain of transmission that would spread MRSA to thirty-three other patients across fifteen months, grievously sickening twenty-seven of them, and killing or contributing to the deaths of seventeen.[28]

Reconstructed, the outbreak looked like this: The man transferred from Texas—call him patient 1—was admitted to a thirty-bed burn unit. Patient 2 was already in the unit, and then became infected. Patient 1 died in early July; a few days after his death, Patient 3 was admitted to the unit, and also became infected in turn. Then came patients 4, 5, 6, 7, and 8, successive admissions to the burn unit who all became infected with MRSA in their debrided wounds. Patient 9 was also admitted to the unit, but then developed MRSA pneumonia so severe that it required a transfer to the ten-bed surgical intensive care unit on a different floor. After patient 9 left the burn unit in October, patients 2, 7, and 8 died; patient 4 had died earlier.

Susceptibility tests early in the outbreak had shown that this MRSA strain was resistant to a number of drugs: methicillin, tetracycline, erythromycin, and clindamycin. In the surgical intensive care unit where patient 9 had been sent, the hospital was conducting a drug trial of an older antibiotic called chloramphenicol. When the MRSA strain passed from patient 9 to patient 10 in an adjacent intensive-care bed, it picked up resistance to chloramphenicol as well. Patient 10 had been in the intensive care unit since September, before patient 9 came up from the burn ward. Because he now had multidrug-resistant MRSA, he would remain in the unit until the following January. While he was there, his MRSA strain would infect fifteen additional patients, and nine of them would die. In January 1981, patient 10 was sent to a rehabilitation ward, where the strain passed to three more patients; two died.

Toward the end of 1980, Harborview tried to shut down the outbreak by putting any infected or colonized patient in isolation. In March 1981, the hospital also closed down the surgical intensive care unit, and additionally opened a new burn unit. Of the outbreak victims, only patient 10 remained in the hospital, in the rehab ward. All the rest were dead or had been discharged. And yet, in May and June, six new infections surfaced in patients in the new burn unit. They had no connection with the previous patients—no family relationships, no crossing paths on wards—and yet their strain was essentially the same.

The only thing the patients could have in common was the health care staff who treated them. Harborview checked everyone who might have had contact with patients in both waves of the outbreak. Out of 182 nurses and physicians and respiratory, occupational, and physical therapists, they found one: a nurse who worked in the new burn unit. She was colonized in her ears, nose, armpits, and groin with the same MRSA strain that had infected the six new burn patients. Further, she had worked in the old surgical intensive care unit where patient 9 had passed the strain to patient 10. And as a student nurse, she had cared for patient 4.

The six newest patients all recovered, but only after lengthy treatment with vancomycin and rifampin. By the time an account of Harborview's epidemic appeared in a medical journal in September 1982, there had been no new cases for a year. But the outbreak could not be declared completely over: patients 10 and 12 were still taking antibiotics, not yet free of the serious infections that they already had been enduring for two years.

Harborview's experience finally persuaded American medicine that MRSA was a burgeoning threat. But action still did not come quickly, and the bacterium did not wait. In 1975, when hospital laboratories around the country tested the staph in their institutions, they found that only 2.4 percent of staph strains were methicillin-resistant. In 1991, in the largest hospitals, the proportion rose to 38 percent.[29] By 2003, in intensive care units, it was 64.4 percent.[30]

By every measure, the situation got dramatically worse as time went on. In 1993, there were fewer than 2,000 hospital stays for MRSA infection in the United States; in 2005, there were 368,600, and each hospital stay cost twice as much and lasted twice as long as one where MRSA was not part of the diagnosis.[31] In 2005, the CDC estimated, 94,360 invasive MRSA infections occurred in the United States, most of them caused by or linked to health care. "Invasive" infections meant only the most serious, including pneumonia, osteomyelitis, and bacteremia; they caused 18,650 deaths that year, more than were caused by AIDS.[32]

Behind every statistic, there lay shattered trust, enormous bills, and irrevocably altered lives.

Jay Sorensen did not have a heart attack; he survived his collapse. His job did not. He had missed so much work that the radio station let him go, on Andrea's birthday, in January 2007. The hole in his abdomen healed in early February. But in April, he began to feel a bulge again. He went back to Nagy, who sent him for a CT scan. As Jay feared, it was another hernia—not in the recent surgical repair, which had held, but in a new weak spot in the muscle, above his original surgical scar. His fifth surgery was scheduled for June. Then, in May, Andrea discovered a lump in her breast. She had a needle biopsy at a local walk-in surgery center and it went smoothly—but two weeks afterward, she woke up on Memorial Day weekend, teeth chattering with chill, and realized she had a fever.

She called her doctor, but he was headed out of town. He sent her to the ER, where they looked at her breast—bright red, swollen, sparking with pain whenever her arm or clothing brushed against it—gave her an IV of pain medication, and drained the infection with a syringe as long as her hand. It took two syringes to get all the pus out. Some was sent to be cultured. The infection was MRSA—but whether she picked it up at the surgical center or from Jay was impossible to know. It took several more doctor's visits to clean and close the abscess, along with an expensive course of linezolid. It left her with a breast so badly scarred that Nagy recommended reconstructive surgery, but her infection and Jay's many recurrences left her too distrustful of health care to attempt it.

Jay's fifth surgery went ahead in June 2007, closing the latest hernia as well as another new weak spot in the muscles that Nagy found. Yet again, the MRSA recurred. That meant another hospital stay and more vancomycin, and this time heavy-duty pain medications that left him in a junkie-like state of withdrawal. Through August and September, cellulitis—a rough red inflammation of the skin that signals infection but

47

produces no pus to culture—bubbled up near the surgical scar. He took oral antibiotics at home to combat it. The new surgical wound was slow to close as well, and a wound-care nurse came to clean the incision and pack and bandage it where the edges had not healed. In December, Jay stepped into the shower to clean up for the nurse's visit, and to soften up the packing so he could remove it and wash the wound before she arrived. He pulled, and felt something give way internally. A piece of packing had been left in the incision; fluid had developed behind it and then worked its way to the outside of his body. It left a channel that could be vulnerable to infection. He went back to Nagy for his sixth surgery, to close the newest wound, in February 2008.

"He has probably the worst wound healing that I have ever encountered," Nagy said. "It's hard to say why; it could be a lingering, subacute infection in the soft tissue, but it's difficult to test for that. Once you cut through those muscles, even when they heal, they will not be as strong."

Throughout 2008, Jay examined his healing incisions anxiously, panicking at the sight of any redness, popping another course of antibiotics whenever his physicians thought something might be blowing up again. In September, his abdomen began to ache and he felt a new lump emerging, close to his groin and below the surgical scars. Whenever he moved or shifted, it bulged. His doctors referred him on to another team of surgeons, in Philadelphia, who had more expertise operating low in the body close to the genitals. On December 30, 2008, Jay had his seventh surgery, a multihour procedure with several surgeons that involved unwinding a loop of intestine that had slipped downward out of his abdomen and was wrapped around one testicle. It took more than one hundred staples to close the incision. It was more than three years since his first infection had turned an uncomplicated surgical recovery into an indefinite ordeal.

Jay and Andrea estimated that, from 2005 to 2008, his insurance spent more than $1 million on his care. But to them the true cost went far beyond that number. It was the irritable bowel syndrome Andrea developed from worrying about Jay and the one hundred pounds she lost as a result. It was the visits with their children and grandchildren that they sacrificed for fear of infecting them. More than anything, it was the irrevocable change in Jay: still big—though nowhere near as big as he had been—but no longer optimistic or confrontational or bold.

"I don't have words for what this has done to my psyche," he said in New Jersey, one day in 2008. "I think of my mortality, a lot. I think of what this has done to my wife. She has posttraumatic stress disorder from this, I'm sure of it. I dread looking down at myself in the shower, wondering if it's going to start again. I think I have a fear that, if I get it another time, none of the antibiotics are going to work."

That fear was not unreasonable. In Chicago, Robert Daum's team had been patiently unwinding the genetic mechanisms that had allowed hospital-acquired MRSA to flourish and community-associated MRSA to emerge. The team's contention that community MRSA was a different entity requiring different control strategies was about to be bolstered— not by scientific agreement, but by a wave of community infections that would sweep the United States.

CHAPTER 4

THE PUZZLE PIECES

Methicillin resistance appeared in England in 1961 and the United States in 1968, but it spread far beyond those two countries. Within ten years of its first identification, MRSA was in Poland, Ethiopia, India, and Vietnam.[1-4] In the 1970s, it was recognized in Switzerland and Denmark.[5,6] In 1979, a MRSA strain invaded hospitals in the Australian state of Victoria, and then spread back to England via an Australian health care worker who took a job there.[7] In 1989, a different strain surfaced in a hospital in Barcelona, and within a year spread to Portugal.[8] In the 1990s, yet another strain appeared in a hospital in Belem, in the northern corner of Brazil, and expanded throughout South America.[9] By 2000, MRSA would be the most commonly identified antibiotic-resistant pathogen in most parts of the world.[10]

Molecular analysis revealed that the pandemic was not uniform; one superstrain of MRSA had not evolved in an unknown place and rolled like a plague across the globe. But neither had MRSA been created fresh each time in each location, out of the near-infinite range of staph strains with their enormous armamentarium of destructive enzymes and toxins. Instead, all of the MRSA strains in the worldwide hospital epidemic fitted into a relatively narrow array of eleven clones, the term for near-identical strains that could be shown to have common ancestors.[11]

They were named for where they first arose geographically: the Iberian clone, the Brazilian clone, the Japanese clone, the Berlin clone, and more. But by no means had they stayed confined to those locations. The breadth of their spread suggested how adaptable and nimble MRSA must be. The Iberian clone surfaced in Barcelona and Madrid, and within less than five years spread through Italy, Belgium, Scotland, and caused a hospital outbreak in New York City.[12] The Brazilian clone spread as far south as Chile and as far east as the Czech Republic.[13]

That speed and reach were stunning, even for a bacterium as ancient and inventive as staph. Investigators took it as proof of the dire impact that the introduction of antibiotics had had on a long-established microbial world. They described it with concern as "accelerated evolution . . . condensed into a remarkably short time frame of four decades."[14] Humans and our staph companions had evolved, slowly and in relative harmony, over the hundreds of thousands of years of hominids' existence and millions of generations of staph. But then antibiotics had forever altered that relationship, and man's ancient microbial companion had responded in its own defense, in ways that had not been foreseen.

Because staph's reactions to antibiotics seemed so unpredictable, it was important to understand how the bacterium had become methicillin-resistant. That was not an easy task. The knowledge came stepwise over more than a decade. In each step, researchers went to the limits of their existing technology, and then discovery stalled until new technologies were developed and refined.

The first clue was the realization that researchers could still detect methicillin after they introduced the drug into cultures of methicillin-resistant staph. With penicillin, that had not happened. The enzyme that broke open the beta-lactam ring, the four-cornered arrangement of atoms at the center of the penicillin molecule, inactivated and destroyed the drug: if you added penicillin to a culture of penicillin-resistant staph and then performed a test to find it, you would not be able to detect the drug. But even the earliest studies of MRSA, conducted as soon as methicillin resistance was uncovered, noted that methicillin was still present at the same time that methicillin-resistant bacteria kept growing.[15,16] The studies had a tone of surprise. Methicillin was based on penicillin, and so researchers expected that the encounter between the bacterium and the daughter drug would proceed in the same manner that the bacterium and the original had. The new drug, like the old drug, should have bound to staph's cell membrane as the bacteria reproduced, weakening the structure so that the organism could not maintain its integrity—but it did not.

The discovery that followed from that clue was that staph had developed a different method for evading the new drug's disruptive effect. Instead of disabling methicillin, the bacterium had worked out a way of ignoring it.

Staph possesses particular proteins on its outer surface; they mark the spot where penicillin locks on to hijack the process by which the cell assembles new membranes as it reproduces. (Penicillin was developed and marketed long before scientists realized that this was the way its molecular chemistry worked; at the time of its discovery, knowing that it killed staph was more important than understanding how.) In 1984, more than two decades after methicillin resistance began to spread, researchers in New York found that the proteins on the surface of MRSA were strikingly different from the ones on drug-sensitive staph. The new proteins simply did not allow the drug molecule to bind. So the drug could not interfere with the cell membrane's synthesis, the staph cells kept growing unaffected, and the impotent methicillin floated in the culture, intact.[17]

The discovery that methicillin did not attach to staph helped to explain something that doctors treating MRSA victims had spotted early on. MRSA was resistant to a wide range of drugs, from the earliest synthetic penicillins to several newer drug families. All of the drug molecules shared that central component, the beta-lactam ring. As it turned out, that ring was the mechanism in the drug molecule that allowed it to latch onto the protein in the cell membrane. Once the proteins changed, none of the drugs containing the ring could bind to MRSA, and therefore none of them were effective.[18,19]

In the two decades since that recognition of MRSA's basic evasive strategy, the bacterium had acquired many additional resistance factors beyond its baseline indifference to beta-lactam antibiotics. But those were present only in the strains that spread in hospitals. Robert Daum's group at the University of Chicago had demonstrated that community-associated cases had different resistance patterns. They were beginning to perceive that community and health care MRSA caused different illnesses as well.

After twenty years of hospital outbreaks, medicine knew what to expect of MRSA. It flourished in health care institutions where patients were immunologically vulnerable. It attacked those who were very young or very old; or whose immune systems were undermined by chronic diseases or long drug regimens for cancer, AIDS, or transplant rejection; or whose natural protective barriers had been interrupted by surgery, trauma, or burns. It caused specific, well-recognized syndromes: endocarditis, life-threatening infections of the valves of the heart; bacteremia, an overwhelming bacterial load in the blood; surgical-wound infections, and infections related to implanted catheters and artificial joints. And because of the number and variety of drugs used in health care institutions, it was resistant not only to penicillin and methicillin and their beta-lactam cousins, but to a vast array of additional classes of drugs.[20]

But the cases that had been found in Chicago, Minnesota, and elsewhere, the ones with no health care ties, did not match those parameters. The victims had not been hospitalized. They were neither newborn nor elderly, and they did not have diabetes or cancer or AIDS. They had pyomyositis, pus-filled abscesses buried deep in muscle; osteomyelitis, infections of bone and bone marrow; pneumonia that ate holes in lung tissue, and sepsis. They were children like Simon Sparrow, son of Everly Macario and James Sparrow, who woke with a shriek on April 16, 2004.[21]

Macario and Sparrow had lived in Chicago for about eight months, brought there by Sparrow's new job as an assistant professor of history at the University of Chicago. It was a rare and precious tenure-track position at a school known for commitment to the liberal arts, so Macario, who had been teaching at Harvard, had been content to let her husband relocate them. Since the move at the start of the 2003 school year, she had been working from home as a consultant, a choice that let her raise seventeen-month-old Simon and his older sister Elena, who was four and a half. Macario's last degree had been a ScD, the PhD-equivalent conferred by the Harvard School of Public Health. Her specialty was public-health campaigns that persuaded people to use condoms and wear bike helmets and eat more vegetables.

"I thought we were done with infectious diseases," she said ruefully. "I thought what I was doing was what we needed to do next to improve our health."

In April 2004, the couple had just filed their taxes and signed the papers for a Hyde Park condo. Sparrow's new job was going very well, and Everly had as much work as she wanted. "Beautiful kids, dream jobs, a great neighborhood," she said, looking back. "We would hold each other and say, 'It's too good to be true.'"

Simon was big for his age and sturdy except for a touch of asthma, with Macario's dark eyes and a mop of red-gold curls from his father. When he woke disoriented and feverish, they thought he might have a cold, or a return of a throat infection and breathing problems that he had been diagnosed with two weeks earlier. Macario had taken him to the emergency room for that and brought him home with a prescription for antibiotics and steroids. It was 7:30 a.m. and Elena had a stomach virus that Macario was already handling, so she let her husband take Simon to the ER this time. A few hours later, with Elena tucked in at home and being minded by their nanny, Macario met Sparrow at the hospital for a hand-off. He was due to drive to Peoria, three hours to the south, to give a speech. Simon was restless, squirming and wanting to be held, but X-rays and all his test results were unremarkable. The ER staff sent Macario home with him.

On the way out of the ER, she noticed that his lips looked blue.

At home, Elena was throwing up, and soon Simon was too, though he had not had anything all day but water. He rested on Macario's lap, and after an hour or so, she noticed that he was laboring to breathe, pushing out his chest and using the little muscles between his ribs to pull air in. And he was cold to the touch, icy cold, not feverish the way he had been that morning. She called a friend who was a pediatrician, held the receiver up to Simon's nose and mouth, and then got back on the line.

"Hang up," her friend said. "Call 911."

An ambulance came quickly, and a knot of physicians met them at the ER and conducted them up to the pediatric ICU. They fired questions at her. Had they been out of the country? Was he a preemie? Did he have any allergies? Macario was shaken by their intensity, and perplexed by the commotion. One of them took her into a little conference room. The doctor had a sheaf of paper, test results that she was rapidly riffling through. "Your child is very, very sick," she said. "He has an infection and we don't know the source. We're going to be here a long time."

But it was not long at all. Simon's heart sped up to 190 beats a minute. His blood pressure sank. His skin darkened with petechiae, pinpoint hemorrhages that showed his blood was clotting abnormally as he sank into shock. Then his skin began to peel away. His lungs filled with fluid. At about 10:00 p.m., as his frantic father reached the ICU, the doctors said that Simon's lung tissue was too ravaged by the infection to work. They recommended he be put on a bypass machine that would take the load off his lungs by shunting the blood out of his body and aerating it mechanically. They tried it. It made no difference. Simon slid fully into septic shock, and one by one his organs failed. At noon the next day, less than twenty-four hours after he left the ER to go home, he was pronounced dead.

Numb and bewildered, Macario and Sparrow agreed to the hospital's request for an autopsy. Two months later, in June 2004, they were asked to a meeting with a raft of researchers. The group waiting for them included Robert Daum, now the chief of pediatric infectious diseases;

Chris Montgomery from critical care; and Susan Boyle-Vavra, PhD, an associate professor and molecular microbiologist who was the co-principal investigator for Daum's team. Simon had had MRSA, the team told them. It was the kind the Chicago team had been tracking for eight years, the virulent community strain that caused life-threatening illness without any warning. It had killed another child, a nine-month-old girl, just about a year before Simon died. She had lingered for six days, but both had had the same severe course: respiratory distress, disordered blood chemistry, organ failure. Autopsies on both showed their lungs were threaded with spots of dead tissue. And there was something else odd: bleeding in their adrenals, small glands above the kidneys that produce hormones governing metabolism and stress response.

That particular combination—fever, petechiae, adrenal hemorrhage, septic shock—had been in medical textbooks for almost one hundred years. It was called the Waterhouse-Friderichsen syndrome, after its discoverers, but it had nothing to do with MRSA. It happened rarely, and usually in the overwhelming assault of bacterial meningitis.[22] No one had ever seen it related to staph before.

Macario and Sparrow could barely absorb it. "I have a PhD in public health, and I had never heard of MRSA," she said. "My brother is a physician, my father is a physician, my parents had done laboratory work on staph. They hadn't heard of it becoming community-acquired. But what was crazy, and still is crazy, is that we have no idea where he got it, and we have no idea why he was susceptible."

Boyle-Vavra, who served as the Chicago team's chief microbiologist, was convinced that the differences between hospital and community MRSA, the altered susceptibility patterns and striking clinical course, had a genetic basis. She wanted to prove the hypothesis true.

Her suspicion was based on a succession of molecular discoveries in the 1980s and 1990s. First had come the recognition that the manufacture of the protein on staph's surface to which methicillin could not bind (its technical name, with typical scientific creativity, was "penicillin binding protein 2a") was governed by a particular gene dubbed *mecA*. That gene was a puzzle, because no one who was expert in staph could say with authority where it had come from. Staph's most important genes had been identified in the 1980s, and *mecA* had not been among them.[23]

The gene appeared to have been acquired from another of the many species of staph, and researchers argued fiercely over which it had come from and how it had been imported. Bacteria have an array of strategies for altering their DNA. There is the one that everyone thinks of, mutation: changing nucleotide by nucleotide in response to some stimulus. But mutation's changes are random, and as a survival strategy it is slow, the genetic equivalent of waiting for all the cherries to line up in a slot machine. Bacteria have evolved faster strategies that allow them to acquire entire DNA segments. They can trade them cell to cell by opening their membranes to each other, a process called conjugation. They can receive them via phages, viruses that infect bacteria, which is called transduction. Or they can pick up, from their environments, free DNA released by other bacteria that have died and lysed or broken open, a process called transformation.[24]

It was not microbiologically possible for *mecA* to have arrived in MRSA alone; it had to have been accompanied by other genes called recombinases that would lever *mecA* into staph's genetic code. The recombinase genes served as guideposts for researchers exploring how the gene *mecA* had arrived. In 2000, Japanese scientists announced that *mecA* had traveled into staph as part of a small segment of DNA.[25] The technical term for such a small segment is a cassette, and the researchers dubbed this one "staphylococcal chromosomal cassette," or SCC, *mec*. SCC*mec* was small enough that it could have migrated into the bacteria via a phage or by several other pathways, but the cassette lacked distinctive genetic elements that would have clued them into its method of transport. That and its origin remained a mystery.

Further work, announced the next year, showed that there were three varieties of SCC*mec*, each containing the *mecA* gene that conferred methicillin resistance.[26] The cassettes were of varying sizes—one was smaller than the first one identified, and the other was larger—because each one contained a few additional genes that conferred resistance to specific antibiotics: tetracycline, erythromycin, and other drugs. Those genetic findings matched what clinicians had learned from outbreaks of hospital MRSA: the strains were always resistant to the beta-lactam antibiotics, and usually resistant to other drugs, but the resistance factors were not universal or uniform.

Since the community strain was resistant only to beta-lactams, Boyle-Vavra wanted to know whether its cassette looked different from the ones that had already been identified. From the time the Chicago group published their first paper, they had been storing MRSA isolates from any patient whose symptoms and history matched the new phenomenon they were seeing. They approached the Japanese researchers who had sequenced the known *mec* cassettes, Teruyo Ito and Keiichi Hiramatsu of Juntendo University in Tokyo, and offered a collaboration. If the two groups worked together, marrying the Chicago trove of isolates and clinical experience to the Tokyo group's expertise in identifying SCC*mec* sequences, they should be able to elucidate any difference in the new strain faster than either group could on its own.

The Tokyo group agreed, and the teams' joint results in 2002 proved Boyle-Vavra's intuition was correct: the community strains were different at the molecular level. MRSA strains found in children such as Simon Sparrow contained a cassette that was much smaller than the others. They possessed the *mecA* gene that conferred beta-lactam resistance, but that was all. Unlike the cassettes found in the hospital strains, there were no genes conferring resistance to additional families of drugs. The Japanese researchers had dubbed the earlier discoveries SCC*mec* type I, type II, and type III. The new one was formally named SCC*mec* type IV.[27]

The smaller size of the newly found cassette immediately suggested a question: was SCC*mec* IV a descendant of one of the larger cassettes, a reduced version that had shed most of its resistance factors but retained beta-lactam resistance at its core? The query was not idle—it was, in fact, the microbiological challenge to the Daum group's contention that community MRSA was fundamentally different. If community strains were really just hospital strains gone mild—escapees from a health care institution that had shed their unneeded resistance factors like a jacket that was too tight—then Daum's entire research enterprise might be based on a false premise.

The Chicago team took isolates from their collection that possessed the newly identified small cassette, and put all of them through three separate tests, including the molecular fingerprint–generating PFGE, to distinguish how alike or different the strains were. It was already known that hospital MRSA strains were not very diverse, but fell into one of those eleven international clone types no matter where in the world they originated.[28] If the Chicago isolates matched those clones too, the case for their hospital origin would be strengthened, and the contention that they were different from hospital staph would be proven wrong.

But it was not wrong. The three sets of tests demonstrated that the genetic makeup of the community strains carrying the small cassette was strikingly different, not only from the major hospital clones, but from each other.[29] Community MRSA strains were not enfeebled hospital purebreds. Instead, they were new hybrids that were being created from scratch in the world outside the hospital. The *mec* cassette was inserting itself over and over into many different drug-sensitive staph strains, creating a diverse pool of community MRSA whose relationships to each other looked less like a family tree and more like a swarm.

Microbiologically, that was interesting. Medically, it was troubling. Staph had always been a variable, formidable bug. If the diversity of the newly recognized community strain meant that staph's longtime virulence was converging with its newfound drug resistance, that could be very bad news indeed. From about 2002, physicians in cities around the United States began suspecting that process was under way.

Outside of Chicago, the idea that MRSA came only from within a hospital—and only occasionally migrated outside it, though never for long—was firmly entrenched, especially at the largest teaching hospitals in Boston, New York, and Baltimore. There was some justification for that. As far back as MRSA's first U.S. emergence at Boston City Hospital, MRSA outbreaks had tended to occur in the largest hospitals that housed the sickest patients. By the time the Chicago and Tokyo teams were undertaking their analysis, academic institutions and their powerful faculties had been handling MRSA for two decades. They did not have it under control, but they felt they understood it. ·

But there were other researchers willing to challenge the orthodoxy of MRSA's origin and behavior. They worked in hospitals with reputations that were more regional than national, and therefore had less stature to risk—places and professors that served their communities ably, but that Harvard faculty might have had difficulty finding on a map.

They were places such as Corpus Christi, Texas, where two physicians counted up the cases of MRSA in children at Driscoll Children's Hospital between 1990 and 2001 and bluntly declared an "exponential increase . . . accounted for by an explosion of [community-acquired] MRSA infections."[30] And Memphis, where staff noticed that MRSA coming into Le Bonheur Children's Medical Center rose by two-thirds between 2001 and 2002.[31] And the Children's Hospital of Minneapolis, where an analysis of lab records revealed that the rate of MRSA cases tripled between 1991 and 2003. Rates of methicillin-sensitive staph at that hospital did not change in those four years. MRSA was not replacing other infections; it was creating additional new ones.[32]

The first group to really sense the scope of the problem, and try to measure it in an organized way, was just as far outside the Harvard-Hopkins axis: Texas Children's Hospital, the very large pediatric training institution for Baylor College of Medicine in Houston. In 2000, the infectious diseases team there, headed by Dr. Sheldon L. Kaplan, recorded every staph infection coming into the hospital between February and November. At first, MRSA looked like it would not be such a challenge. Though the drug-resistant strain was common, it was not as common as drug-sensitive staph; underlining staph's long-standing reputation for virulence, drug-sensitive staph actually caused more serious infections more frequently.[33] The next year, though, the hospital started surveillance to catch every MRSA case, paying special attention to infections that originated outside the hospital. With the numbers that surveillance project returned, they could see the growing problem of community-associated MRSA unfold almost in real time. By 2002, the community strain was the most frequent cause of empyema, pus-filled cavities between the lung and chest wall.[34] By 2004, it was more likely than drug-sensitive staph to cause deep muscle and bone infections, and children whose illness was due to MRSA had longer fevers, longer hospital stays, and a much higher rate of complications.[35] By 2005, it was four times more likely to cause invasive pneumonia.[36]

Thanks to their comprehensive surveillance, the Houston group could discern several troubling trends. First, the amount of illness caused by the community strain was growing, in raw numbers and as a percentage of all staph infections. Second, while the vast majority of the illnesses were minor—skin and soft-tissue infections, things that could be treated with a clinic or emergency room visit—the proportion that needed overnight care in a hospital was increasing. And third, when the community strain caused serious illness, it was very serious.[37] The Houston team found what the Chicago team was also finding: severe bone infections and dense walled-off abscesses buried deep in big muscles.[38,39]

But they also found syndromes that the Chicago group had never seen, proving the prediction that community MRSA, created out of multiple staph strains, would have diverse and unpredictable effects. Some of the child patients at Texas Children's had not only previously recognized symptoms—fever, bone infections, bacteria in the blood—but also large, dense blood clots that blocked veins and caused tissue death wherever blood flow could not reach. Venous thromboses, VTs for short, sometimes happen in adults who are confined to bed or a sitting position without being able to move about; they are a known danger on long airplane flights. But they are vanishingly rare in children, or at least they had been, until the community strain of MRSA arrived. Between 1999 and 2004, the Baylor team treated nine kids with MRSA and VTs, most of them with clots buried so deep in their legs that they were not discovered until the children underwent MRIs.[40]

Shortly afterward, a team at Children's Medical Center in Dallas reported the same thing: deep clots in children with community MRSA infections, accompanying osteomyelitis and usually close to the infected bone. In some children, the infection showered septic emboli, blood clots that also contained infection, into their lungs, brains, and hearts.[41] Then another group in Dallas—at Parkland, the enormous public hospital that older Americans know as the place where President John F. Kennedy was declared dead—found yet another very serious syndrome. They discovered that the community strain of MRSA was causing eye infections so large and deep that the infection and pressure could destroy a child's sight if not handled immediately.[42] And at Le Bonheur hospital in Memphis, one of the institutions that had first signaled the community strain's assault on children, doctors began recognizing cases of septic arthritis, difficult-to-diagnose infections that could permanently damage knee or hip joints and ruin for life a child's ability to run and play.[43]

As these cases racked up, the few researchers tracking them began to have doubts about the contention that the diverse new community strains had arisen from common drug-sensitive staph. Drug-sensitive staph had never been as consistently virulent as these new strains were turning out to be. They wondered whether in its evolution the community strain had picked up some additional element, something that further enhanced MRSA's ability to cause serious infections.

In France, a research group was pursuing the possibility that one of staph's many toxins could be responsible. The toxin was called Panton-Valentine leukocidin, PVL for short: Panton-Valentine for the two researchers who identified it in 1932, and leukocidin for its ability to lyse white blood cells.[44] PVL was not a common staph component; in France, it showed up in only 5 percent of the isolates sent to the national reference laboratory from hospitals across the country.[45] But the scientists had had their eye on PVL for a while, because it seemed to play a role in infections that involved tissue destruction such as pneumonia, though not in the kind of bloodstream and heart infections that occurred in hospital patients. The French scientists had drilled down into their national database and retrieved and compared the records of patients infected with staph strains that did and did not express PVL. The PVL-positive strains were drug-sensitive except for penicillin resistance, and except for being able to produce PVL, they were genetically different, not clones. They had identical effects on their victims, though: the PVL-positive patients were much sicker. They were more likely to die, they died much faster—in an average of four days compared to twenty-five—and autopsies revealed that their lung tissue was destroyed, ulcerated, and bloody and with its architecture blown apart.[46]

That description sounded so like the earliest cases of community MRSA that the Hiramatsu group in Tokyo was intrigued. After identifying the *mec* cassettes, they had gone on to sequence the complete genome of two MRSA isolates from hospital patients in Japan; simultaneously, two other hospital MRSA strains had been sequenced by other researchers in England. To see how they differed from the strains in community outbreaks, the Tokyo group obtained one of the first community isolates: the methicillin-resistant staph that had killed a sixteen-month-old girl from a North Dakota Indian reservation in 1998, one of the first cases reported by the CDC.

The girl had been catastrophically ill. She died two hours after her panicked family got her to a local hospital, and when her body was autopsied, it revealed that she had not only devastating pneumonia and sepsis, but MRSA infections scattered through her internal organs. When the Tokyo group analyzed the strain that killed her, which had been given the name MW2, they found good reasons why: its genome was studded with virulence factors, eighteen exotoxins and enterotoxins that would mount a wide-ranging attack and stimulate an overwhelming immune response. One of the toxins was the French group's culprit, PVL.[47]

The *mec* cassettes had not been identified when the CDC reported the girl's death in 1999. But with the tools they had achieved, the Tokyo group could now confirm that it was a community strain that killed her—it possessed the distinctively small SCC*mec* type IV. PVL was present also, but it was not in the cassette. The two genes that cause it to be produced were in a different place in MRSA's genetic code, accompanied by markers that showed they had been transported into MRSA at some point by a phage, a virus that infects bacteria. Those genes had not been found in any of the sequenced hospital strains. PVL and its unique destructive abilities were only a property of the community strain.

The recognition inspired other researchers to look for the toxin too. Community-associated MRSA, identified by the type IV cassette, had spread around the world as quickly as hospital strains had spread two decades earlier, and analyses found the PVL gene combination in widely different strains in France, Switzerland, Australia, New Zealand, and Western Samoa.[48] The Chicago team identified it in Simon Sparrow, also infected with MW2, after his death. The Texas Children's Hospital researchers identified it in high proportions—but not 100 percent—of the children they had treated for severe pneumonia, pyomyositis, osteomyelitis, and deep venous thromboses. And in a lovely piece of retrospective detection, British researchers retrieved preserved samples of staph 80/81 from Australia, England, and the United States, and sequenced them. They found that almost all of them contained the genes for PVL, a discovery that might explain that lost strain's extraordinary ability to cause disease.[49,50]

Not everyone agreed that PVL was the source of community MRSA's virulence. Some researchers argued that it could not be responsible because, in animal tests, strains in which the PVL genes were artificially deleted caused abscesses and sepsis just as severe as those caused by PVL-producing strains.[51] Others said that heretofore unknown virulence factors, unrecognizable until the molecular tools to identify them were achieved, must actually be responsible for the cellular destruction ascribed to PVL.[52] Still others countered that, while staph's disease-causing ability might be due to many virulence factors acting in concert and not to any single toxin, nothing but PVL could cause the necrotizing pneumonia that was community MRSA's most virulent expression.[53]

Boyle-Vavra, the Chicago team's chief of microbiology, was enormously intrigued by the toxin. When she looked in the collection of MRSA isolates that they had been assembling since 2003, PVL genes were in all the community-associated strains. Yet there was no good explanation so far for why that should be so. PVL was so common, across so many community MRSA strains around the world, that some extraordinary selective pressure that was consistent across the planet must have been keeping it there. Antibiotic use was not a sufficient explanation; PVL does not play a role in resistance, and it is present in drug-sensitive as well as drug-resistant strains. Yet the PVL genes and the type IV *mec* cassette were found together so consistently that their joint presence—

SCC*mec* type IV, PVL-positive—was becoming medical shorthand for "community-associated MRSA strain." They remained, for the time being, one of staph's many mysteries.

"I am willing to accept it as a marker," Boyle-Vavra said in 2008, in her lab in the old Wyler hospital pavilion, converted now to offices, where Daum and Herold had spotted the first community MRSA cases twelve years before. "If it is not directly responsible for any of the ways in which MRSA is virulent, it is closely enough associated that it shows us something is going on."[54]

The database at University of Chicago was unusually valuable—not just because it was so comprehensive, but because it was so rare. Boyle-Vavra was speaking almost exactly four decades after the drug-resistant strain made its first U.S. appearance. In all that time, there had been no national data collection, for hospital strains or for the newer community ones. No one, anywhere in the country, could say how many MRSA infections actually occurred in the United States, or which strains caused them, or whom those strains attacked.

Large hospitals with good microbiology labs tracked their own local epidemics, and research universities with an interest in the bug kept whatever archives they could fund out of their grant proposals. The CDC, the federal agency responsible for public health surveillance, kept an eye on MRSA through several databases: hospital infections through voluntary reports from three hundred to five hundred hospitals, a fraction of the 5,700 hospitals in the country, and community infections through voluntary reports from a selection of emergency rooms, outpatient clinics, and doctors' offices. After 2004, it also tracked infections both in the community and in hospitals through the Active Bacterial Core Surveillance System, a network of hospital labs and health departments in ten states, covering about 38 million people, about 13 percent of the population.

None of the data collection was required. MRSA has never been what public-health professionals call a reportable disease, for which physicians must notify health departments and health departments must notify the CDC. None of the databases were comprehensive, and few of them were linked. It was up to interested researchers to notice when something happened. Along with Chicago and Houston, there was another such group in San Francisco. And that was fortunate, because MRSA's attributes and its victims were about to change dramatically again.

CHAPTER 5

THE BIGGEST THING
SINCE AIDS

When she looked back, Kepa Askenasy had trouble making sense of her harrowing experience. Her business undermined, months of work lost, relationships postponed, and her trust in medicine gone forever. All for something that started, in the chilly San Francisco summer of 2007, as a small white spot on her left eyelid.

The spot was the width of a pencil lead, but it hurt out of all proportion to its size. As an architect, Askenasy crinkled her eyelids hundreds of times a day to squint at photographs, CAD renderings, the details of three-dimensional models. Now, every time she peered at a drawing, she felt a lancing pain that worsened as the day went on, and a flush and pressure that spread above and below her eye.

Her regular doctor was out of town, so she went to a clinic near AT&T Park, a few blocks away from her home at the northern edge of Potrero Hill. One of the physicians on duty took her vital signs, went over her history—unremarkable for a healthy fifty-one-year-old, no allergies, no addictions, no surgery save a hysterectomy—and then looked at her eye.

"Oh, you've got the urban epidemic," the doctor said immediately. "That's MRSA; everyone's got it." She prescribed doxycycline, explaining carefully that it was an old drug that dated back to the 1970s but still worked reliably against the resistant bug. She sent Askenasy home with a recommendation to use hot compresses until the infection subsided or burst, predicting that she'd be back to normal in seven to ten days.

Askenasy took her first dose of antibiotics as soon as she returned to her home, a compound comprising an old house and several sheds that she had converted into a sleek private retreat. The property included an elegant studio for herself, a rental unit for income, a garage and a garden room where her cats could retreat when the fog forced them indoors. It would have been unaffordable anywhere else in town, but the neighborhood, hemmed in by freeways and shipyards, had been slow to gentrify. Rougher and less affluent than the rest of San Francisco, Potrero Hill is thinly populated after dark, and quiet except for the rumble of traffic heading to the Oakland Bay Bridge. Whenever Askenasy unlocked the unremarkable door that led her from the sidewalk into the walled garden, she felt as though she were stepping into a Zen refuge from the city.

MRSA was about to make her retreat a prison.[1]

Within a day of starting the doxycycline prescription, Askenasy was desperately ill. She had a fever, chills, light-headedness, leg cramps, and abdominal pain so intense that she could not fold at the waist to sit down. She lay on the sofa with her laptop propped awkwardly on her ribs, and Googled the symptoms for doxycycline allergy. Frightened by what she found, she called her doctor's office, hoping to be switched to another drug. Her physician, who had returned from vacation, listened to Askenasy's symptoms with concern. She was not convinced the drug was to blame, but she agreed to authorize a different prescription—a sulfonamide, a drug class that has been on the market even longer than penicillin. When the new drug also had no impact, Askenasy was switched to a third drug: ciprofloxacin, a fluoroquinolone that has been on the market since the late 1980s. "You should be better soon," her doctor said. "Cipro kills everything."

But Askenasy did not get better. Her abdominal ache intensified so much that she couldn't take deep breaths. Her joints throbbed. When another course of Cipro produced no improvement, her doctor started sending her for tests—complete blood work, urinalysis, an abdominal CT scan, kidney ultrasound, and a spinal MRI. The tests found nothing to explain her constant pain and weakness or her intermittent fever. In late August, she began breaking out in pustules, first on her right calf and thigh, then on her groin and arms, and finally on her face. They were all the size of a pencil eraser, and they all followed the same course. A pustule would rise for a few days, come to a head, rupture, then sink, leaving a dark mark on her skin. She began to feel like a leper, and forbade friends with children from coming to visit. She also gave up on working; the pain made it hard to concentrate, and lying flat made it too difficult to conduct research or draw. Eventually, she left her house only for doctors' appointments. She underwent an endoscopy of her esophagus, stomach, and small intestine, and then a colonoscopy to examine the large intestine. Neither test showed anything significant.

With the help of the Web, Askenasy became convinced that her initial MRSA infection—presumed on the basis of the first doctor's clinical instinct, but never actually tested for or confirmed by culture—had never been eliminated. But among the many physicians helping her, that idea got little traction. "They said MRSA was a skin infection, not an internal infection," she said. "And they told me I had already taken the right drugs."

The pustules had subsided. Frustrated, Askenasy stopped taking any antibiotics, hoping they would return and give her some evidence her physicians would be forced to heed. The small bumps did not return, but in mid-September, she developed a single large abscess on one buttock. It swelled to the size of the palm of her hand, broke open, and drained. When she showed it to the next physician she visited, a gastroenterologist whom she was referred to in late September, he prescribed a course of two drugs: Septra, and rifampin. The first was a combination of two more older antibiotics; the second was yet another older drug that provoked staph resistance when used on its own, but was believed to enhance the killing power of other drugs.[2]

For the first time in three months, she got some relief. The abscess healed, and the nausea and fevers abated. But within days of her last drug dose, the whole panoply of symptoms returned: weariness, abdominal ache, joint pain, fever. Aggravated and sick, she confined herself to home again, alone except for a few friends whom she hoped would not be at risk of infection.

"You know what this reminds me of?" one visitor asked her one day. "Do you remember when people got so sick from tampons, in the 1980s? Toxic shock, they called it. I wonder if that's what you have."

Molecular confirmation in 2002 that community-associated MRSA was a distinct entity from the hospital strain underscored a problem that had been building for some time: what to call the proliferating range of resistant staph strains. Over forty years, multiple naming schemes had been used. Depending on a researcher's personal preference, strains might be identified by phage types, place name, *mec* cassette, or PVL expression. Some of the nomenclature was specific to certain countries; the British called their hospital strains EMRSA, with E indicating "epidemic," and the Canadians called all their strains C-MRSA—with C standing for

Canadian, not community. Many researchers separated strains into categories using the results of a gene-sequencing technique called multilocus sequence typing or MLST, which distinguishes among small genetic differences. The MLST results were designated "ST" for sequence type; the United States' hospital strains, for instance, fell into ST 5. Some idiosyncratic names lingered in the medical literature, causing confusion long past their original usefulness—such as "type G" and MW2, names that were given separately by Chicago and Minneapolis researchers to the same first community-associated strains.

Sensing imminent confusion, the CDC proposed a new and different naming scheme based on the distinctive patterns generated by PFGE, the lab technique that produced "molecular fingerprints" that let researchers differentiate among strains. PFGE uses an enzyme to snip bacterial DNA into segments; the segments are pulled through a gel sheet by electrical current to generate the bar code–like fingerprint. The technique is relatively inexpensive compared to gene sequencing, and the equipment and knowledge needed to perform it are widely distributed in hospitals and research universities and in the laboratories of most state health departments. The DNA fingerprints that resulted could easily be digitized and shared electronically to compare isolates found in different locations, something the CDC had been doing for foodborne disease since the early 1990s.[3]

Scientists at the agency collected staph isolates from CDC investigations and hospital microbiology labs, and from other countries' national laboratories for reference. They assembled 957 isolates altogether, originating from the United States, Europe, and Asia between 1995 and 2003. When they put the isolates through PFGE, they discovered that the strains grouped naturally, by their visual similarities, into eight clusters.[4] They called the clusters USA100 through USA800.

The strains that fell into the clusters dubbed USA100, 200, 500, 600, and 800 were the familiar old multidrug-resistant hospital strains, carrying SCC*mec* types I through III. Of them, USA100 was by far the most common, making it the main type seen in hospitals. USA300 and 400, on the other hand, originated in community outbreaks; they carried SCC*mec* type IV and made PVL. They were resistant only to beta-lactam antibiotics and to erythromycin. The USA400 cluster included the strain that Robert Daum and his colleagues had called "type G" and Tim Naimi and the Minnesota researchers had called MW2, the one that had caused the overwhelming and fatal illnesses in children from Chicago,

Minnesota, and North Dakota in the late 1990s. USA300 was a newer arrival, first recorded in 2000 in a patient in San Francisco, that seemed to be responsible mostly for skin and soft-tissue infections. Also in the mix were USA700, a small cluster that included both community and hospital outbreaks, and seventeen isolates that did not fit into the eight-group scheme at all.[5]

By sorting MRSA strains into readily identifiable, named groups, the CDC gave researchers and physicians a tool for counting and tracking local outbreaks. Fortuitously, the system was developed just in time to document what seemed to be a vast increase in community-associated MRSA infections caused by the upstart strain, USA300.

The first upticks appeared in the CDC's data from a late-2000 investigation of skin infections in a Mississippi state prison and a very large outbreak of boils and abscesses in the Los Angeles County Jail.[6] Jails and prisons are unique medical ecosystems. Many of the people who end up there are already in shaky health from drug use, marginal living conditions, or poverty. Plus, the conditions inside jails and prisons—crowded, with little access to health care—exacerbate the spread of infection. So it was possible to dismiss those first observed outbreaks as peculiar to penitentiaries and not significant for overall public health.

But then physicians around the country began reporting surprising increases in the rate of MRSA infections. They were occurring in people without prior ties to health care facilities. They were found during emergency room visits, or so soon after hospital admission that they had to have started outside the hospital—so they were community infections by definition. They were the kind of minor infections—cellulitis, boils, and abscesses—that would normally have been caused by everyday drug-sensitive staph, but were turning out to be MRSA. The new infections signaled that the community strain—USA400 and now USA300 as well—could not only cause the rare, invasive syndromes identified by the Chicago and Houston researchers. Those might have been triggered by some undetected genetic quirk. The strain could also cause a range of illnesses in perfectly healthy children and adults with no connection to the health care system and no impairments in their immune systems—in short, with no reason to believe they were at any risk at all.

In 2003, a study at Grady Memorial Hospital in Atlanta found that 72 percent of all staph infections at the hospital and its outpatient clinics were caused by MRSA, and almost all of them were USA300.[7] In early

2004, 75 percent of staph infections in the emergency room of Alameda County Medical Center in Oakland, California, were MRSA, and 87 percent of them were USA300.[8]

USA300 seemed to advance with such striking speed, from nonexistent to dominant in less than four years, that teaching hospitals with longstanding microbiology databases went back through their records to document its spread and to make sure they had not missed something. Their results all confirmed that USA300 was both new and fast-moving. At Olive View–UCLA Medical Center, MRSA rose from 29 percent of staph infections in 2002 to 64 percent in 2004. All of the 2004 isolates put through PFGE were USA300.[9] In the emergency department of the Baltimore VA Medical Center, MRSA as a percent of staph infections went up tenfold between 2001 and 2005—from 4 percent to 42 percent— and almost all of the infections were caused by USA300. The number of patients seeking help for skin and soft-tissue infections tripled.[10] In August 2004, ERs in eleven cities analyzed all the skin and soft-tissue infections they treated that month. Averaged among them, the amount of staph that was methicillin-resistant was 59 percent—though that ranged from 74 percent in Kansas City down to only 15 percent in New York, demonstrating that community-associated MRSA was not moving across the country in a uniform way. Still, on average, 97 percent of the MRSA isolates were USA300. In addition, 74 percent of them were a single identical strain, dubbed USA300-0114. For a bacterium that usually was extraordinarily diverse, this was unheard-of.[11]

In addition to all those skin and superficial infections, USA300 was also displaying an ability to cause dangerous, invasive infections just as the earlier, more diverse community strains had done. In 2005, physicians from Harbor-UCLA Medical Center in Los Angeles somberly reported that over sixteen months, USA300 had caused fourteen cases of necrotizing fasciitis, "flesh-eating disease," a rare, hard-to-detect, and imminently life-threatening infection of the fibrous covering that separates muscles from the fat and skin above them. Necrotizing fasciitis is most commonly caused by group A *Streptococcus*. The fact that the disease was now also caused by staph was a deeply disturbing discovery to the doctors who treated the fourteen victims. To survive flesh-eating disease, patients need not only emergency surgery, but immediate treatment with IV broad-spectrum antibiotics—yet the resistance factors that MRSA had already developed made such treatment less likely to succeed.[12]

The warnings of USA300's advance would have special meaning for doctors in San Francisco, a city with the most comprehensive MRSA surveillance in the United States, where USA300's resistance profile was changing.

San Francisco's interest in MRSA came about because of an impending embarrassment. In 1998, San Francisco General Hospital was preparing for a visit by representatives of the Joint Commission, the powerful nonprofit organization that evaluates whether hospitals are in compliance with federal regulations and thus eligible for millions of dollars in Medicare reimbursement. As part of the portfolio "the General" would submit to the reviewers, Dr. Henry F. "Chip" Chambers, chief of infectious disease, asked the hospital epidemiologist to gather statistics on the percentage of staph infections in the hospital that were methicillin-resistant.[13]

"Because MRSA was a hospital bug," Chambers explained. "And if the rate's going up, you've probably been doing a bad job."

Unfortunately for Chambers, the epidemiologist's report indicated that the rates had been going up.

Chambers has worked at University of California–San Francisco, the med school that supplies faculty to the General, for his entire career, since arriving as an intern in 1977. Nonetheless, his voice retains the lilt of central Kentucky, where he was born and went to college. In moments of stress, his slight accent strengthens to a twang. Describing the MRSA numbers, the twang became very strong. "The data showed that, in the ten years I'd worked with the infection-control program, there had been an inexorable and unabated rise in MRSA rates," he said. "So we agreed I was doing a terrible job. Then we broke the data out based on parameters that would let you make a guess whether it was from the community or in the hospital, and that showed us that the increase could almost entirely be attributed to people coming in from the outside."

Needing to count up MRSA cases for the Joint Commission visit at just the moment when community MRSA in San Francisco was becoming obvious enough to be detected was a serendipitous accident. It was one in a string of fortunate coincidences that all contributed to MRSA's elucidation. The first signal of community-associated staph surfaced in Chicago just as Robert Daum's group was looking for new research projects to tackle. The fatal cases in Minnesota and North Dakota occurred at about the time that Tim Naimi—an in-demand CDC disease detective who could have worked in almost any state he chose—picked a posting in a state health department that possessed both generous funding and a relaxed attitude about working across jurisdictional lines.

But in San Francisco, serendipity seemed to be working overtime. The city had already created a publicly funded system for delivering medical care: the Community Health Network, comprising San Francisco General, the rehabilitation hospital Laguna Honda, the health services in the five city jails, and multiple clinics serving a number of neighborhoods and also special populations such as minorities, street youth, and transgendered adults. In addition, because the city possessed a sizable population of homeless and marginally housed people who were vulnerable to minor wounds and infections, the General had created a separate clinic, the Integrated Soft Tissue Infection Service or ISIS, just to keep them from clogging the emergency room. Then, because all those institutions—the hospitals, the clinics, the jails, and ISIS—shared one electronic network, patients' medical records plus microbiology results from the network's central laboratory were already collected in one place, available to any medical researcher who might be interested in them.

It was a final piece of good fortune that the General's staff included a UCSF faculty member who found that huge database particularly interesting. Francoise Perdreau-Remington, PhD, an elegant and brisk microbiologist and professor of medicine, founded the Molecular Epidemiology Research Laboratory in 1995 on the hospital campus. Ever since, she had been collecting drug-resistant bacteria from across the community network, about 95 percent of all the staph strains found in San Francisco's hospitals, clinics, and jails. Crucially for a project exploring the natural history of a new bacterium, all of the isolates arrived with abundant additional detail—not just the patients' medical histories and treatment, but also information about past illnesses or conditions that would predispose them to the bug, and even the neighborhoods where they lived.

"A database like this, it lets you play detective," Perdreau-Remington said. "What that means is following, on a daily or at least weekly basis, all the data you receive from the lab, and trying to make sense out of it—analyzing not just the genetics and the resistance pattern, but the patients, their locations, their lives."

Poring over the database, the UCSF researchers could pinpoint USA300's arrival in San Francisco in late 2000. They could see it ticking upward in the community health network's patients: from 145 isolates per year in 1996 to 341 in 2000 to 1,240 in 2004. In 2002, it accounted for two-thirds of the MRSA infections in San Francisco's five jails, and one-third of those in clinic and hospital patients at San Francisco General. By 2004, the proportion was more than half.[14-16]

"We started seeing what we called the S clone, what the CDC would call USA300," Perdreau-Remington said in her small, orderly lab. "It was a pattern on the PFGE gels, different from all the others. Usually you put fifteen samples on a gel, and there would be perhaps three or four that looked the same. The next gel, perhaps eight. And eventually we would pull gels and they would be the same pattern, all the way across."

Even with such a vast database—2,000 staph strains by 2002, and more than 10,000 by 2008—there were questions about the new strain's behavior that Perdreau-Remington and her lab colleagues couldn't answer with the data they had at hand. People of all classes come to the General, especially for emergency care—but most of its regular patients and those of its clinics are economically disadvantaged or socially marginalized and may lack the living conditions, mental status, or income to take competent care of themselves. Comprehensive as it was, the community network's data was telling the researchers only about the behavior of USA300 within a narrow slice of San Francisco's enormous economic and social range. They needed to see the larger picture.

To broaden their reach, they recruited help from other local hospitals. Seven of the eight other medical centers in San Francisco, and one just across the county line, agreed that for twelve consecutive months, starting sometime in 2004, they would collect a bacterial sample from every patient with a suspected MRSA infection, whether the infection began in the hospital or outside. The hospitals collected 3,985 isolates. Because some people had been sick more than once, their infections were subtracted from the total, yielding 3,826 cases in one year in a city of 719,000 people. Those 3,826 were then subdivided into cases that started in hospitals and ones whose symptoms started at home or so soon after hospital admission that they could not have been infected there. After sorting the strains microbiologically, the researchers found that among the community-onset cases, USA300 accounted for 78.5 percent.

That high percentage confirmed that what was true for patients at the General and the city's health network was also true throughout San Francisco. For additional verification, the UCSF team performed one more piece of math. When epidemiologists compare the occurrence of a disease in different areas, they use a rate, a word that in disease detection specifically means the ratio of cases of disease to every 1,000 or 100,000 members of the local population. Using the case count they had gathered from the eight hospitals, the UCSF team calculated that the rate of hospital-acquired MRSA in San Francisco was 31 per 100,000. But the rate of community-associated MRSA was 316 per 100,000—more than ten times as high.[17]

In ten years of watching community MRSA emerge across the country, no one had ever seen numbers like that. A few years earlier, CDC researchers had tried similar calculations, collecting MRSA data from eleven Baltimore hospitals, eight counties around Atlanta, and six rural and six urban hospitals in Minnesota. In their findings, community MRSA accounted for 8 percent of all MRSA diagnoses in Baltimore, 12 percent in Minnesota, and 20 percent in Atlanta.[18] In San Francisco, the proportion was 80 percent.

The majority of the San Francisco cases were cellulitis, abscesses, and boils—manifestations that disease detectives lumped together as "skin and soft tissue infections" and tended to consider minor problems. But they were not minor to the patients themselves. Physicians in San Francisco were seeing skin infections that started as small pimple-like eruptions—patients almost universally suspected a spider bite—which then blew up into deep abscesses or opened into weeping wounds with a black center of dead tissue. Four-fifths of the patients coming into ERs with skin infections needed at least minor surgery, and at least one in six required hospitalization for more significant surgery, as well as antibiotics that could only be delivered by IV.[19]

Chip Chambers, the General's hospital epidemiologist who had rung the first alarm about community MRSA in San Francisco, paid special attention to the seemingly minor infections. They suggested to him that San Francisco's MRSA problem was greater than anyone appreciated.

"Just because they are more common, they represent a huge reservoir of possible transmission," he said in his tiny office behind a warren of cubicles. "That's a spawning ground for more severe disease, in a population sense. When you increase the burden of disease, you increase the incidence of any complications that are a consequence of that disease. And as an individual, if you are colonized, you are more likely to have an infection, and if you have a minor infection like a skin infection, that predisposes you to something more serious like a bloodstream infection."

The UCSF team knew that MRSA colonization was more common in San Francisco: in a 2000 survey, completed before USA300 first appeared, they estimated that 2.8 percent of the San Francisco population were asymptomatic MRSA carriers, three times higher than the national average.[20] The greater the pool of colonization, the greater the chance that MRSA would adapt in some unforeseen manner. With so many bacterial strains in battle with many immune systems, subjected to antibiotic pressure and other environmental assaults, it was likely something unpredictable would happen. USA300 was genetically stable; it was highly transmissible; and it was virulent, causing necrotizing fasciitis and other deadly manifestations such as abscesses of the brain.[21] It was a better colonizer of humans than the hospital strains, and in the lab it grew faster. Its limited amount of resistance was all that kept it from rivaling or surpassing hospital strains of MRSA as a grave public health threat.

In her lab, Perdreau-Remington could see evidence that US300 was becoming more resistant, and that worried the UCSF group even more than its broad distribution across their city did. They sequenced a sample of USA300, choosing a USA300–0114 isolate from their collection, taken from the wrist abscess of a thirty-six-year-old man whose infection had taken an unusually long time to heal. They named it FPR3757— the initials for Perdreau-Remington, and the number because it was the 3,757th bacterial isolate they had collected and stored.

Since Daum's first paper, it had been understood that community-associated MRSA strains were resistant to fewer drugs than the hospital strains. Indeed, the narrow resistance profile and the small size of the cassette SCC*mec* type IV—small because it had not acquired the genetic clutter of additional resistance factors—were the accepted ways to identify a community strain. But by the time the UCSF team started studying USA300, community MRSA had become resistant not only to the beta-lactam family of antibiotics, but to some additional drugs as well, chiefly a drug family called macrolides that includes the common antibiotic erythromycin. The genetic maneuvers by which MRSA gained erythromycin resistance showed how devious the bacterium could be, because they were both new and redundant. One created tiny pumps that pushed the drug out of the bacterial cell, and another changed the part of the bacterium where the drug molecule would have locked on. That second change was dismaying because it also conferred resistance to another common drug called clindamycin that physicians used when erythromycin did not work. The clindamycin resistance was not absolute; in fact, it occurred only when clindamycin was administered to someone whose MRSA was already erythromycin-resistant. It was as though the bacterium anticipated what doctors would try next, and blocked their next move before they made it.[22]

Losing erythromycin and the other macrolides had been problematic. Potentially losing clindamycin was a terrible blow. It was a reliable, inexpensive drug that could be given by mouth, meaning that it could be taken at home. It penetrated efficiently into the skin, making it a good choice to treat skin infections, and it had a unique ability to shut down staph's production of its many destructive bacterial toxins. The amount of community-strain staph that possessed inducible clindamycin resistance was still low, however, and there were still plenty of other drug options for outpatient treatment: the tetracyclines; the fluoroquinolones; the sulfa drugs; and linezolid or Zyvox, a brand-new and therefore much more expensive drug that could also be taken orally.

The sequencing of FPR3757, though, demonstrated how fast treatment options were vanishing.[23] It revealed a mutation that created resistance to the fluoroquinolone drug ciprofloxacin—the antibiotic that had failed to eradicate Kepa Askenasy's infection. That was new. So were genes that conferred complete resistance to clindamycin and also never-before-seen resistance to tetracycline; to another group of drugs called streptogramins that included some formulas that had just come on the market; and finally to the drug mupirocin—the main ingredient in the antibiotic gel used to decolonize patients' noses.

In addition to the new resistance factors, FPR3757 also possessed mobile genetic elements, small sections of DNA that had been transported in from other bacteria by phages. One DNA section contained genes that produced proteins that could help the bacteria spread and defend itself against white blood cells.[24] Another section contained a sequence imported from *S. epidermidis*, a benign colonizing strain, which enhanced USA300's ability to survive on skin.[25] A third new element held genes that allow staph to make two toxins that were new to community MRSA strains. Like the gene that produces the toxin responsible for toxic shock syndrome, an epidemic of fatal staph disease linked to tampons that swept the United States in 1980, these new genes made "superantigens," proteins that spark an overwhelming immune response. That was especially worrying, because other staph researchers, not in Chambers's group, believed that superantigens are an underappreciated cause of staph diseases resembling toxic shock syndrome. Those would be especially hard to diagnose and treat, because they do not resemble MRSA's classic skin or bone or joint infections. One group in Minnesota had recently analyzed the case of a man who died of necrotizing pneumonia and an overwhelming breakdown of the circulatory system, who had been infected with a superantigen-producing, community MRSA strain.[26,27]

The results of the sequencing helped explain why USA300 had spread so widely, squeezed out other strains so completely, and caused such a range of serious illnesses among otherwise healthy people. It revealed a so-far unmatched combination of transmissibility, virulence, and resistance. And though the UCSF researchers had never met or treated Kepa Askenasy, the results also suggested why her physicians were having such a hard time curing her.

By the late fall of 2007, Askenasy was sick of being sick, of not socializing or earning an income, of shutting herself off from the world. She started seeing a nutritionist, and started exercising again, riding her bike painfully up the San Francisco hills. She felt a little better, enough to begin calling clients and soliciting work. Though she didn't feel normal, she was so relieved to be able to sit up and walk around that normal seemed like a luxury she could forego.

In December, as she was slicing foamboard for an architectural model, she dropped her X-acto knife, and cut a large gash in her thigh. Disgusted with her clumsiness, she hauled herself to the emergency room for stitches. While taking her history, one of the residents asked if she had had any illnesses. Hoping for a reaction, Askenasy told him she'd had MRSA.

"Oh yeah," the doctor replied nonchalantly, "we've all had that."

Three days later, the wound was inflamed and oozing under the stitches. Askenasy went back to the ER. They diagnosed MRSA by sight and sent a culture for confirmation. The report came back: "Heavy staphylococcus aureus (methicillin resistant)." If the MRSA infection from earlier in the year had ever been eradicated, it had now returned.

She recited her drug history, explaining that the Septra and rifampin combination had worked for a while before, so they gave her more of it, two weeks' worth. A few days later, she had an idea. She had been planning to take a trip to Spain as a reward for enduring a half-year of illness and isolation. Thinking about the trip, she realized it presented an opportunity to be prescribed a second round of drugs in case her symptoms recurred while she was away. She cycled back to her doctor and got the additional round approved. But she did not wait for a recurrence. Instead, as soon as she finished the first course, she started on the second, effectively prescribing for herself a course of antibiotics that was much longer than a doctor would have recommended. By the time she took the last dose, she was in Barcelona waiting for the all-night street party that marks New Year's Eve. The MRSA, as far as she could tell, was gone. The extra-long course of antibiotics knocked it out of her system, though it left her with a mild paranoia about infections, and deep anger over having been ill for so long.

"I had probably $100,000 worth of tests, and $20 worth of medicine was what fixed me," she said. "I should have been better in two weeks."

The infectious diseases staff at San Francisco General understood epidemics. In the early days of AIDS in the United States, the hospital was ground zero. The earliest HIV cases were recorded in Los Angeles and New York City in 1981, but it was the General, at the heart of the city that was the heart of gay America, where the wave of illness crashed down. Chip Chambers had been chief resident in medicine in 1981, and he had never forgotten that extraordinary time: the shifting definitions of what was wrong, the uncertainty over how best to take care of people, the astonishing, climbing count of the ill.

Watching the advance of MRSA, he was reminded of those days. The organism was less frequently deadly, but it was far more widespread, and possessed an almost willful ability to evade the body's defenses and medicine's control.

"Other than AIDS, this is the biggest thing in the past thirty years," he said. "In incidence, in cost, in impact on health care, in how you approach patients. There are so many things where you must assume MRSA could be causing them: Earaches in kids. Sinus infections. Pneumonia. The big three sites for infection are respiratory tract, urinary tract, and skin and soft-tissue, and suddenly half of those skin and soft-tissue infections are resistant. Your workhorse antibiotics, gone. And if you put someone on an ineffective antibiotic, you may make things worse, because it furthers resistance and it prolongs the period when they are infectious."

In its rapid emergence, USA300 challenged most of what researchers presumed they knew about community strains of staph. Hospital strains were thought to be uniquely multidrug-resistant because of conditions in the institutions that bred them, yet USA300 had become multidrug-resistant in the outside world. It was also believed that hospital strains caused the most serious diseases: in late 2007, CDC scientists published a paper that investigated the origin of the most invasive cases of MRSA during 2005, and calculated that 86 percent were due to hospital-associated strains.[28] Yet as early as 2002, the UCSF team had uncovered that 14 percent of the MRSA infections that occurred in patients after they were hospitalized, and therefore should have been caused by hospital strains, were actually USA300. By the end of 2005, the proportion had become 43 percent.[29,30]

USA300 had also surprised them because, while staph overall had always been known for its diversity, the single MRSA strain of USA300–0114 had crowded out so many others. Now, though, it was surprising them a further time: USA300–0114 was losing its dominance, and USA300 was becoming more diverse. In January 2008, Perdreau-Remington's team estimated that there were twenty-five variants of USA300 circulating just at UCSF. As that strain differentiated, it was reasonable to assume that some of its clones would be less resistant and less virulent; ineluctably, though, some would be more so.[31] In September 2008, CDC researchers demonstrated that that was happening. In a few USA300 isolates, they uncovered additional resistance factors: to tetracycline drugs, to gentamicin, and to trimethoprim-sulfamethoxazole, also known as Bactrim or Septra—the bedrock combination antibiotic of community MRSA care. The additional resistances occurred rarely, but where they did, they made USA300 almost as drug-resistant as a hospital strain of MRSA. When the researchers probed further, they discovered where those new resistance factors had come from. USA300 had acquired them from USA100, the main hospital strain.[32]

MRSA strains that had always been understood to cause infections only outside hospitals were causing infections inside them. Strains that had always been understood to have less resistance than the hospital variety suddenly boasted just as much. The terms that had described MRSA for decades were no longer useful, because the bacterium itself was no longer behaving as it had for all those years. The hospital and community epidemics were converging.

CHAPTER 6

THE FLOOD RISING

Because the United States lacked national surveillance, Chip Chambers and the UCSF researchers could not be sure whether San Francisco's extraordinary rates of community-associated MRSA were unique across the country, or whether physicians there were simply better at tracking the bug. Scattered reports, though, suggested the problem was much more widespread than most doctors knew. At the VA Medical Center in Washington, D.C., the number of people seeking help for MRSA skin infections increased fourfold between 2001 and 2005.[1] In central Massachusetts, the number of community MRSA cases per month went up ten times between 2003 and 2006.[2] When the Tennessee Department of Health tracked MRSA cases in that state between 2000 and 2004, they found that the number of skin and soft-tissue infections caused by MRSA had increased a staggering sixteen times.[3]

The only problem with those reports and others like them was that very few people were hearing them. They were presented at medical meetings, sometimes as talks but usually as "posters," which is just what it sounds like: a large sheet of paper pinned to a board in a warehouse-sized room lined with hundreds such boards. Posters are medical meetings' response to the reality that there is far more research conducted every year than there are slots in the agenda of a specialty society's annual gathering. They are the way that new, small, or geographically limited research findings begin to make their way into public discourse. Frequently, though, they go no further. Only a few posters at any meeting are chosen for publication in major medical journals, which are the main method by which medicine builds its collective base of knowledge, and also the main route by which new research gains the attention of the mainstream media.

Starting in 2005, at least a dozen research teams presented evidence of the advance of community-associated MRSA at medical meetings. Their findings were important signals that went almost completely unheard.

In San Francisco, Chip Chambers and his team felt much the same way Robert Daum and his colleagues in Chicago had almost a decade earlier—as though they were describing a phenomenon that almost no one else could see.

"First, there's the skepticism of, Oh, this really doesn't exist," Chambers said, sounding more stressed, and therefore twangy, than usual. "Then there's, Well, it exists—but it's only a problem in Chicago, or in San Francisco, or in that one jail. But if you go to Campbellsville, Kentucky, where I grew up, they have community MRSA—because when I go back there I talk to the doctors there, and they see it. It's just that there's no surveillance for it. So if you don't count the problem, how do you know how big it is?"[4]

In 2007, one of Chambers's associates recognized a way to capture the problem's size. Dr. Adam Hersh, a UCSF research fellow in pediatrics, suggested they plumb federal databases that collect the details of visits to physicians' offices, outpatient clinics, and emergency rooms—collectively, the majority of medical care performed in the United States that does not require an overnight stay in a hospital. The databases include codes that indicate the patients' diagnoses, as well as information about the drugs they received. As a lens for looking at MRSA's nationwide impact, the databases weren't perfect. The International Classification of Diseases, the codes that are used in medical records, stratify diagnoses by body site and by symptoms, not by disease-causing organism—so in all of medical record keeping, there is no single widely used code indicating MRSA infection.[5] However, because USA300 had become the single biggest cause of skin infections brought to emergency rooms, Hersh thought it seemed reasonable to pull the records for any visits for skin infections, abscesses, and cellulitis. To those in the medical community who had not been tracking MRSA as vigorously as the UCSF team, the results were stunning. Between 1997 and 2005, the rate of skin infections had almost doubled; in emergency rooms and among children, it had tripled. MRSA was producing 14.2 million visits to doctor's offices, clinics, and ERs every year.[6]

When he drilled down into the data, Hersh found even more troubling news. Prescriptions for antibiotics that could effectively attack MRSA were increasing much more slowly than MRSA was rising. Three-fourths of the prescriptions written for skin infections were beta-lactams, the drugs MRSA had been impervious to for more than forty years.[7]

Since early 2006, the CDC had been working to get the message to doctors that prescribing patterns must change. The agency convened an expert panel of thirty-four researchers who devised an algorithm to guide prescribing. Unless a patient needed hospitalization, doctors were to use clindamycin, trimethoprim-sulfamethoxazole, and the tetracyclines, while staying aware that some resistance to that last class was emerging. They were warned to be cautious about the use of fluoroquinolones and macrolides, because resistance to them was growing. And they were strongly urged to reserve the new drug linezolid for emergencies.[8] The CDC published this advice on its website and distributed it to health departments and medical societies. But the guidelines were unenforced recommendations: doctors could choose not to follow them. Some never knew the recommendations existed.

Steve McNees grew up on the beach in California and spent twenty years walking streets in Salt Lake City as a mailman. In February 2007, at the age of sixty, he was reaping the consequences of a lifelong disdain of hats and sunscreen: he had a half-dollar-sized basal-cell carcinoma on the back of his head and his dermatologist wanted to take it off.[9]

His in-office surgery went smoothly, but just to be sure the incision stayed clean, his doctor gave him a prescription for two weeks of the beta-lactam antibiotic Keflex. The wound took longer to heal than McNees expected, and after a few weeks, the skin near the incision bubbled up into little pustules that spread all over his head. His physician switched him to three weeks of amoxicillin, another beta-lactam. In late April, with the hole in his head almost closed—but, in McNees's words, "not quite right"—he discovered a sore spot on his butt. Suspecting an ingrown hair, his doctor gave him azithromycin, a macrolide.

The sore did not subside, though; it opened into a weeping wound deep in the cleft of his buttocks. The doctor prescribed Avelox, a fluoroquinolone, and referred McNees to a colorectal surgeon to make sure his colon wasn't involved. Although the surgeon thought the abscess was only superficial, he switched McNees to yet another drug, ciprofloxacin, which didn't help, either. The sore grew to egg-sized, and the hole in it was the size of a quarter. To absorb the seepage, McNees started shoving a dishtowel down his pants.

In June, the hole in his butt seemed to be closing, but he noticed an angry spot on his lower eyelid. He went back to his primary-care doctor, who prescribed more amoxicillin and sent him to an eye specialist to make sure the skin cancer hadn't recurred. Ruling out cancer, the specialist said it was probably just an infection. To be safe, though, the eye doctor scheduled a biopsy, and switched the prescription to Keflex again. Meanwhile, McNees could feel another boil rising on his butt. When he went for a checkup of the first sore, he showed the second to the colorectal surgeon. Later, when it blew up to the size of a tangerine, the doctor recommended it be surgically cleaned out. To prepare, he gave McNees another prescription for Avelox.

The surgical cleaning took place in July, more than five months after the incision on McNees's head started the merry-go-round of infections and prescriptions. Searching his symptoms on the Internet, McNees and his wife Sue came across descriptions of MRSA. It sounded like what he had, so they insisted the surgeon take a culture of the abscess before operating. The results came back positive for MRSA, which meant that, of the seven rounds of antibiotics McNees had taken since February, not one had been appropriate for his infection.

The culture result also meant that McNees was referred to yet another doctor, an infectious-disease specialist. Because the abscess was already open and draining, the doctor did not prescribe more antibiotics, but he did swab McNees's nose to check for MRSA colonization. When that test also came back positive, he prescribed a decolonization regimen: a week-long course of Bactroban, an ointment containing the antibiotic mupirocin, dabbed into the nose. McNees followed the regimen, but spent most of his energy dealing with the fist-sized open wound that the surgery had left on his butt. Even with weekly visits to a wound-care clinic, it was closing very slowly, and it hurt when he walked, or sat on anything that wasn't cushioned, such as a bench or a church pew.

As Labor Day approached, he and Sue decided to take a break from the city heat, and headed up to Garden City, a remote lakefront town near Utah's border with Idaho. The day before they left, Sue noticed what looked like a bug bite on her shoulder. By the time they arrived, her "bite" had puffed up to the size of an egg. The next morning, it was the size of a soup can, and from shoulder to elbow her arm was flaming red. They found a small medical clinic and were barely in the door when the physician on duty spotted her arm. "I would swear that's MRSA," he said. "Has anybody else in your house had an abscess?"

82

In Salt Lake City, they had been fighting for months to get doctors even to consider the possibility of MRSA, and here, in the middle of nowhere, a doctor had named it walking in the door. Dumbfounded, Sue pointed at Steve.

The doctor, David Tullis, checked her over, lanced the swelling, and recommended drugs for both of them: trimethoprim-sulfamethoxazole for Steve, and linezolid for Sue because she is allergic to sulfa drugs. It was the first time in eight months that Steve had been given a systemic antibiotic with the knowledge that he had MRSA. He began to get better almost right away.

Scientific attention was turning back to staph colonization, the silent carriage of MRSA strains on the skin, and to the hope that altering someone's personal staph population might reduce their chance of infection.[10] The idea was especially alluring for the community staph epidemic, because the percent of the population colonized and the number struggling with recurrent infections were both rising.

For decades, it was estimated that 30 percent of the population carried staph asymptomatically, some transiently and some for months and years. Of these carriers, about 1 percent had MRSA.[11] Between two studies done in 2002 and 2004, though, the proportion of the population carrying MRSA doubled, from 0.8 percent to 1.5 percent. That might still seem like a very small number, but 1.5 percent of the U.S. population in 2004 equaled 4.38 million people.[12] And other studies found much higher proportions in smaller groups: 3 percent in the soldiers in one unit in Texas, 9 percent among healthy children getting regular checkups at a medical center in Nashville, 16 percent among people receiving regular dialysis at a New York City hospital.[13-15] Even those counts may have been underestimates, because other research indicated that SCC*mec* type IV MRSA, and especially USA300, tended to be carried in places on the body—the armpit, throat, vagina, and rectum—that were not usually tested.[16-18] At the same time, regional data showed more people seemed to be experiencing repeat infections, sometimes very serious ones. Estimates of how many MRSA victims suffered recurrences ranged from 10 percent to 29 percent.[19]

Decolonization had worked for Eichenwald and Shinefield, the researchers who eradicated staph 80/81 from New Yorkers in the early 1960s. They gave just enough antibiotics to families afflicted with S. aureus 80/81 to eradicate the bug from their systems. But they had never planned for their patients to be permanently cleared of staph—only to create a window in which the benign staph strain, 502a, could take hold. That was in an earlier era, soon after the development of many antibiotics and before the emergence of MRSA. Now the burden of antibiotic resistance around the world was so great that physicians and epidemiologists were more conservative about antibiotic use, for fear of stimulating new resistance. Attention turned instead to chasing the bug from its hiding place in the nose. British researchers had tried this on hospital patients as early as 1961 using an antiseptic ointment, but it had no impact on staph carriage or the infections that followed.[20] The idea was so alluring that the stratagem was tried again and again over decades—but most of the trials were so different in design from previous ones that there was little to agree on. Still, it became clear that decolonization was worth the risk in hospital patients who were at particular risk of serious infections and could be closely monitored. Its effects worked for about a week, long enough to protect a surgical patient from accidental self-infection or from passing the bacteria to others.[21] It was much less certain that decolonization worked in the outside world, and the CDC refused to recommend internal or external antibiotic regimens until researchers could show better evidence.[22]

The lack of CDC approval did not stop patients who were enduring multiple rounds of infection from trying antibiotics, antibacterial soaps, and stringent cleaning regimens that were recommended by their doctors or discussed on the Internet. For some people at least the regimens seemed to work, but they were challenging to follow. As a Nebraska family would find, it took extraordinary commitment to do them long enough to see results.

Mollie Logan barely noticed the small red bump that rose on her inner thigh in the late summer of 2004. Her first child, Isabella, was just fifteen days old, and Logan and her husband Brian were too happy, and too sleep-deprived, to pay attention to what looked like a spider bite. The couple grew up together in Elkhorn, Nebraska, on the western edge of Omaha, and married when they were both twenty. They wanted children, but Brian had enlisted in the National Guard to pay for college, and after graduation he was sent to Bosnia. Soon after he returned in late 2003, Mollie got pregnant with Isabella.[23]

Within a few days, the small pimple on Logan's thigh expanded into a hot red band as wide as her palm, and stretched most of the way around

84

her leg. Worried that she might pass something on to the baby, she saw her primary-care physician, who drained the swelling and gave her a mild antibiotic that would let her keep breast-feeding. Thinking the episode was over, she hurried home to Brian and their child—but that night, the entire right side of Isabella's chest swelled and flushed dark red. When Logan picked her up to nurse, she could feel the heat of a rising fever.

The young couple headed for the emergency room, where doctors told them somberly that Isabella's fever was 103, dangerously high for a newborn and a clear sign that her immature immune system was battling a serious infection. For twenty-four hours, they gave Isabella IV fluids and Tylenol, but her fever and the swelling were unaffected. On the second day, they switched her to vancomycin, and the fever began to go down. In the meantime, culture results that had been ordered in the ER came back, establishing that both Mollie and Isabella had MRSA. The abscess on Isabella's flushed, misshapen chest had not shrunk with the drugs, and the doctors reluctantly recommended surgery to clean it out.

"They told me, we only have one chance to get this right," Logan said, weeping at the memory. "It was the toughest thing ever, to hand her over."

The doctors gave Mollie oral antibiotics—though she and her daughter had the same strain, her infection was less serious—and sent Isabella for surgery while the family camped out in the waiting room. The baby sailed through, her swelling subsided quickly, and ten days later, Mollie and Brian happily brought her home. They chalked the episode up to bad luck, a brief early disruption in a family life that would be perfect from now on.

Fifteen months later, Mollie found a hard, red, rapidly growing pimple on her own right breast. *Oh no*, she thought. *I know what this is.* Her infection had recurred. Her primary care physicians referred her to an infectious disease practice, where the staff not only cultured Mollie's infection—confirming it was MRSA again—but asked that Brian and Isabella come in for tests as well. The new physicians swabbed everyone's noses, and MRSA's other potential hiding spots elsewhere on their bodies, checking to see if anyone in the family was colonized and unknowingly reinfecting the rest. Brian was clear. Mollie, despite her new infection, was too. Isabella, however, was colonized—not in her nose, but in her rectum.

In addition to twenty days of antibiotics for Mollie, the doctors asked the family to undertake decolonization in order to chase any undetected bacteria from their skin and their home. Three times a day, the Logans painted the inside of their nostrils with Bactroban. Every day when they showered, they washed their bodies with Hibiclens, a hospital-grade soap containing an antiseptic that lingers on the skin. After they showered and any time they used the toilet, they wiped down the bathroom with Lysol. Every day, they laundered the bed sheets and all of the towels in the hottest water with a cup of bleach. Regularly, they sponged the kitchen with wipes containing bleach. They did the entire regimen for thirty days.

At first, decolonization seemed to work well But just one month after the regimen was over, Mollie found a third boil, so low in her groin that she thought it was an ingrown hair until she felt the familiar soreness and heat. She was frustrated and frightened. She wondered, *How long can I take these drugs without the MRSA becoming resistant to them also? And what happens when I run out of drugs that work?*

When she went back to the infectious-disease practice, she was given her longest course of antibiotics: thirty days. The family underwent a second thirty-day decolonization effort as well. By the time it was over, they were tired beyond bearing of sticky noses and dry skin, of endless loads of laundry, and of the ever-present smell of bleach.

The second round broke the bug's hold at last. After the month of antibiotics and decolonization ended, Mollie's doctors tested her three times, checking both her nose and rectum. She was clear, and in May 2006 they declared her MRSA-free. Brian never developed an infection. Isabella suffered no recurrent infections after her first frightening episode, but she remained a carrier for many months afterward. The experience left the family hyper-aware of hygiene. They made a habit of showering once a week with Hibiclens, and slathered Bactroban on every cut and childhood scrape, always aware that they could contract the bug again at any time.

Doctors didn't understand why USA300 recurred so unpredictably. The strain could bedevil one person for months before the right combination of antibiotics—or Bactroban, bleach, and drugs—knocked it out, or it could jump from one family member to another, with devastating results if any of them were already in poor health.[24] It seemed clear that the bug was hiding out somewhere while patients and their family members were treated; the question was, where?

To find the answer, scientists needed to define "family" a little more broadly.

Dr. Carlo Vitale had been a specialty veterinary dermatologist for a dozen years in San Francisco, a city so wealthy that a specialty veterinary dermatologist never lacked for cases, but he'd never seen a sick cat quite like the one that arrived in his office in 2004.[25]

The cat was male and three years old, an intelligent, affectionate short-hair. He had been rescued from a shelter by a well-off technology executive whose passion was saving cats. He had crusty, oozing sores on his flanks, legs, and tail that wouldn't heal and made him unadoptable. As a city facility with limited money and no room for a sick cat that might stay indefinitely, the shelter put him on the kill list, but the cat rescuer saved him. For a year she took him to vets and got him steroid creams and antibiotics, but nothing worked. As a last-ditch try, she brought him to the tertiary-care hospital where Vitale works, in a sleek converted warehouse in the Mission.

Persistent bacterial infections in cats are so unusual that Vitale thought the case merited further exploration, so he swabbed the wounds and sent the fluid off to be cultured. The lab results surprised him. The cat had drug-resistant staph, which was why the treatments for its ulcers had made no difference. But the staph was not *S. intermedius*, a species that infects dogs, cats, and other companion animals. It was MRSA.

In the long literature of staph infections, there was only one recorded case of a cat carrying MRSA. Back in 1988, an outbreak in a geriatric hospital in England was traced back to a cat that lived there as a shared pet for the residents.[26] The likelihood of MRSA in this cat was so low that Vitale assumed the lab had made a mistake. He sent the isolate to a friend, a university researcher with access to high-level diagnostic equipment. That test revealed the cat was carrying not just MRSA, but USA300. It was the first time the new strain had been found in a cat. The clone that was taking over the country had found an additional niche to live in.[27]

Vitale wondered where the strain had come from. In the 1988 episode, staph had passed from a pet to a human. *Was it possible,* he wondered, *that the bug had gone the other way this time?* He asked the owner to let him take swabs of her nose and the cat's nose. Both harbored USA300. Groping, he asked her whether she had ever had a staph infection. Yes, just in the past year, she said; she'd had three bouts of pneumonia, and a painful, weeping inflammation of her sweat glands called hidradenitis. But she had been to the doctor and taken antibiotics, and all the infections had gone away—so she could not have passed the bacterium on to her pet.

Vitale wasn't so sure.

"My theory is, she contracted MRSA, they treated her with antibiotics and she believed she was cured," he explained. "But the swabs show she's still colonized. The cat had allergies, and it had these ulcers. So it picks up MRSA from the owner's hands, or the owner's nose, and becomes colonized. And then it washes the ulcers with its tongue, and it becomes infected too. That's what we think happened. But we'll never prove it."

The episode ended badly for the unlucky little cat. Though his ulcers closed after courses of antibiotics, he never really bounced back, but stayed chronically ill. A year later, he died. The owner, who remained healthy, let Vitale perform a necropsy. He found the little animal's body riddled internally with spots of dying tissue and walled-off collections of pus, profound evidence of how lethal USA300 could be to the newest species in which it had found a home.

The necropsy testified to the power and virulence of USA300, but it also raised another question: if this potent new strain could colonize and sicken household animals, how much would animals assist in its spread?

Among vets, there was already some knowledge about health-care associated MRSA strains finding a home in dogs. In 1994, the bug spread through the ICU of a Welsh hospital. The conduit into the hospital was two nurses, a married couple, who unknowingly gave it to their patients; but the source of the infection was the nurses' dog, which had a drippy eye infection.[28] In 2001, a St. Louis man and his wife who were both already in bad health—he had a diabetes-related foot amputation, and she was also diabetic and had a transplanted kidney—struggled for almost a year with serious infections. First his amputated stump became infected; then she developed cellulitis twice; then his stump became infected again. Eventually their doctors traced the source to their adored eighteen-month-old Dalmatian, which had no signs of illness, but which the couple admitted always licked their noses and routinely slept with them.[29]

Those episodes and the story of Vitale's patient suggested that a human could pass MRSA to a pet, and that the animal could harbor the bug long enough to outlast the human's course of antibiotics and pass the infection back to its owner. That had discomfiting implications. Two out of three U.S. households have either a dog or a cat or both, and frequently more than one.[30] Pet owners are exposed to their pets for more hours at a time, and are more intimate with them, than they are to most of the humans who might pass MRSA to them. But it soon became clear that

MRSA did not just pass back and forth within households. Either a pet or a human could spark a chain of cross-species transmission that infected other humans and animals as well.

Because concern about the possibility was building among veterinarians, the University of Liverpool's Small Animal Hospital checked all of its staff and all of the dogs and cats it was treating for staph. They found almost no carriers: no cats, one dog, and three of the twenty-seven hospital workers. But they also uncovered a previously unrecognized chain of MRSA infections: a dog with a joint infection in January 2004; a second dog with pneumonia in March; a third dog with a wound infection in June; and a veterinarian and a nurse who were colonized with the same human, hospital-associated MRSA strain that was in the three dogs.[31, 32] There was no way to know whether dog or human was responsible for bringing the bug into the hospital in the first place, but clearly it had been there for months at least, pingponging between animals and humans and possibly passing out of the hospital to the dogs' owners and the veterinarians' families as well.

If MRSA could travel from an animal to a human and on to a second animal, it probably could go from a human to an animal and on to a different human as well. Vitale suspected this kind of transmission was possible, and not long after the sick little cat, he confirmed it. A woman brought in a healthy-looking yellow Labrador. The dog, relaxed and friendly, ambled into the hospital wearing a retriever's goofy grin and wagging a heavy tail that knocked brochures off the tables; the woman, though, was in tears. Her family doctor had sent her, she said. Vitale had to help them, or they would have to get rid of the dog, and they loved the dog.

The entire family had been enduring a MRSA outbreak. The woman, her husband, and their toddler daughter had had deep painful abscesses, with the toddler's the worst. The outbreak started when they left the girl with a babysitter; they later remembered she had complained that she'd broken out in boils after getting a tattoo. The family went to the doctor, and had the bacterium identified and took the correct drugs. They thought the outbreak was over, but their daughter kept breaking out in fresh boils all over her buttocks and back. The dog slept on her bed at night, and their doctor was convinced the dog was the source of the recurring infection.

Vitale swabbed the dog's nose and sent the swab to the lab. The dog had MRSA, and it was USA300.

"So what do I do?" Vitale said. "We're still just researching this. There are no protocols to tell us whether the dog should be isolated, what the best way is to decolonize an animal, what are the right drugs to use. What do we tell the owners? What do we tell the owners' physicians? We just don't know."

Vitale was willing to try what seemed to work in people—after all, the dog had a human strain—but there was no research to confirm his intuition, or back him up if the attempt failed to work. A less aggressive option would be to send the dog to a kennel for a few weeks, in hopes the colonization would burn out on its own—if an affordable kennel could be found that would take a dog suspected of carrying an infectious disease. But no one has good evidence, from experiments or even concentrated observation, of how long a dog carries MRSA before the bug departs, so it was possible that the owner would spend thousands of dollars kenneling the Lab, but find afterward that the dog was still colonized. In the end, she and Vitale opted to try the human regimen: for two weeks, twice a day, they squirted Bactroban up the dog's nose. It worked. The toddler's abscesses healed and did not recur. When they checked the dog again, he was clear.

The realization that MRSA could pass back and forth between humans and animals was especially disturbing to veterinarians. At first, they felt at risk. Several surveys in 2005 and 2006 demonstrated that vets had much higher rates of colonization than the 1 percent or so in the general public: 18 percent in a London animal hospital, 10 percent in a Canadian large-animal hospital, and 7 percent of the veterinarians and 12 percent of the technicians attending a veterinary-medical conference in the United States.[33-35] It was reasonable for veterinarians to fear infection; there is a long history of them catching diseases from animals they care for. Some have been minor inconveniences, like foodborne illnesses such as salmonella passed in animals' poop. But others can be dire: when a shipment of Gambian giant rats spread monkeypox—a disease related to smallpox—across the Midwest, vets made up almost half of the victims.[36] And animal-transmitted diseases can be fatal. In 2003, in the midst of international panic over avian influenza spreading from Asia, a Dutch vet died while investigating an outbreak of avian flu in chickens.[37]

After further surveys within the veterinary community, however, it became clear that the risk could also flow the other way. Perhaps veterinarians were endangering the animals they took care of, rather than solely being endangered by them.

Research by one of the CDC's disease detectives, a public health veterinarian named Dr. Jennifer Wright, revealed how vets might acquire diseases and also pass them on. In 2005, she mailed anonymous questionnaires to members of the American Veterinary Medical Association, asking about the ways that vets behaved around animals they examined. The responses she got from 1,842 veterinarians were disheartening, and a bit embarrassing. Almost half didn't always wash their hands between examining an animal and eating, and more than half didn't wash them when going from one animal patient to another. Five out of six ate in areas where animals were examined, and only one out of four used protective equipment—gloves, gowns, and eye shields—when standards of practice indicated they should have.[38]

Still, even when veterinary staff was thoughtful, and cared deeply about the animals they were treating, MRSA could slip by their precautions, as the San Diego Zoo learned from one of its elephants, called Baby.

The elephant wasn't actually named "Baby"—that and "little girl" were her keepers' nicknames for her. The little elephant, born November 28, 2007, at the zoo's Wild Animal Park, was not supposed to be named for months after her birth.[39] First there would have to be consultations with the government of Swaziland, which had allowed the seven adults in the herd to come to San Diego. The small group had had a sad life. They were born in South Africa's Kruger National Park in 1991, and their parents were shot three years later when game officials wanted to reduce stress on the park's ecology. The juveniles were relocated to a park in Swaziland, where their own lives were threatened because their foraging impinged on grazing areas for other endangered species. In 2003, the San Diego Zoo and Tampa's Lowry Park Zoo negotiated a deal to bring the herd to the United States. The process was tumultuous, with three court challenges from animal rights' groups and a two-month suspension of the zoo's import permit, but a bull and six females arrived at the park in August.[40]

The members of the herd were the first wild African elephants to be added to an American zoo in two decades. In court, the zoos had argued that they were needed for the genetic diversity that would keep the species going in captivity as its home range vanished. So the arrivals of baby elephants were important events. There was Vus'musi in 2004, Khosi in 2006, and Impunga and Phakamile in 2007.[41] The newest baby was highly anticipated as well.

She was supposed to join the herd within a few days of her birth, but things did not go as planned.[42] Her mother, whose name was Lungile, had birth complications and developed an infection. She took care of the 192-pound newborn, nursing her and touching her all over with her trunk—but after a few weeks, the elephant keepers noticed that the baby was suckling for unusually long periods, yet losing weight. With the infection taking most of Lungile's energy, they concluded her milk must be of low quality. On Christmas Eve, they decided to hand-feed and hand-rear the calf themselves.

This was a huge decision. Hand-rearing a baby elephant has been managed in only a few zoos, and the Wild Animal Park had never tried it before. It was not just a matter of teaching the little animal to drink from a bottle. Elephants are very intelligent and extremely social. For the baby to become a normal adult, she would need to get as much play and stimulation from her human minders as she would have received from her aunts and cousins in the herd.

The zoo's veterinarians and keepers closed off a section of the elephant barn with hay bales, making rooms where the baby could sleep and roam, and where the humans could keep supplies and records. They built up her "bedroom" with linens and large pillows and stuffed animals to play with, and scattered her exercise area with toys and balls and things that would pique her curiosity. At least one person at a time kept her company around the clock, stroking and grooming her, blowing gently into her trunk, and bottle-feeding her—baby elephants are supposed to drink two and a half gallons of formula a day. They took her out in the exercise yard for walks and bedded down with her when she rested. When they stretched out on the sleeping area, she slept in their laps.

"We'd take her out to play in the sun, and she'd follow us with an energetic, happy trot," Jeff Andrews, the elephants' animal-care manager and chief of the keepers, recalled. "When she heard us next door, she'd pop her head over the hay bales to see what we were up to. She was the cutest little thing in the whole wide world."

But she was not taking well to the bottle. She was still losing weight and her eyes stood out from her face. To give her extra nutrition, the keepers put her on IV fluids, trundling a stand behind her as they walked her around the barn. In early January 2008, they noticed a rash on her right ear, an area of scratches and small, painful-looking pustules. The rash spread to her neck, where they had made an incision to allow an IV, then to her left leg and elbow. Soon after, zoo veterinarian Dr. Nadine Lamberski noticed a pinpoint pustule on one of her own knuckles.

When she mentioned it to Andrews, he said he had one too. Several of the other keepers did as well.

They worried that the baby, in her depleted state, might have picked up an infection and passed it to her human caretakers. They cultured the elephant's sores for bacteria, and Andrews volunteered to have his checked as well. They were shocked at the result. The sores were caused by MRSA, and the strain was USA300.

In the long history of elephants in captivity, no zoo had ever found MRSA in any of them. There was none in the rest of the San Diego herd, either. It seemed clear that the infection hadn't traveled from the elephant to her humans. It was much more likely that one of them had given it to her.

Andrews and the keepers were upset, and they could not help but feel guilty. Elephants are long-lived even in captivity, and people succeed as keepers because they are willing to spend decades building relationships with the animals and gaining their trust and friendship.[43] The idea that one of them had passed a disease to the youngest, most vulnerable member of the herd was disturbing. To suspect that they had contributed to her deteriorating health was worse.

The zoo soon realized that MRSA must have gone from a colonized human to the little elephant and then back to the other humans who were feeding her, drawing blood from her, cuddling her, and helping her walk. Out of the fifty-five members of the elephant staff, twenty had skin lesions. Some of the lesions, like Lamberski's and Andrews's, were so minor they cleared up on their own. Three caretakers with more serious infections took antibiotics, though none were hospitalized.[44,45]

The little elephant was given antibiotics as well, and her MRSA lesions cleared up too. But the infection was an additional assault on a system made fragile by malnutrition, diarrhea, and other bacterial infections. On February 4, 2008, surrounded by Andrews and her keepers, she was euthanized. She still had not received a name.

MRSA had been expanding its range and its influence, affecting first hospital patients and then ordinary community residents; first humans and then animals; first the ill, and then the apparently healthy with no known vulnerabilities at all. It was about to make another dramatic shift, and demonstrate how quickly it could be lethal to children whose parents had no idea they were at risk.

CHAPTER 7

GONE WITHOUT WARNING

It was New Year's Eve 2005, but Scott and Katie Smith had little interest in celebrating.

Fourteen-month-old Bryce, their only child, had been struggling with a cold since two days before Christmas. Since then, they'd gone back and forth by phone with his pediatrician, trying to catch each other during the busy holiday season. The doctor advised Tylenol, Advil, Nyquil, and patience; he seemed to think they were just anxious new parents. They *were* anxious—Bryce's breathing was harsh and his fever would not break. They thought he was sicker than his symptoms showed.

Just after midnight on New Year's Eve, Katie's anxiety spiked. Normally, Bryce, who had her dark-blond hair and Scott's brown eyes, had a permanent faint tan thanks to Scott's Colombian heritage, but tonight he was yellow, dull and dark like a fading bruise. His belly puffed out and he grunted with effort when he breathed. Frightened, Katie and Scott rolled the limp toddler in a light blanket and tore off through the wide, empty streets northeast of San Diego toward Rady Children's Hospital.

When they ran into the emergency room, before they could even speak, the triage nurse's head jerked up.

"Is that your son?" she asked. "Is that your son breathing like that?"

Panting, Katie nodded. The nurse scooped Bryce into her arms. She banged the button that opened the doors to the ER, and pulled her walkie-talkie one-handed out of her pocket. "Come with me," she said.

In the back of the ER, what seemed like a dozen people pounced on Bryce at once, pulling off his clothing, checking his blood pressure, lifting his eyelids, and flicking a flashlight across his eyes. A nurse slid a small clip with a bright red light inside it onto Bryce's big toe to measure his pulse rate and the oxygen in his blood. Beyond the doctors' shoulders, Katie saw the read-out. From watching TV programs she remembered that pulse-ox, a measure of how much oxygen was getting into Bryce's blood, was supposed to go to 100. Anything less than 95 was dangerous. She watched the numbers tick upward. They stopped at 82.

Words bounced up from the health care workers huddled around Bryce: *Pneumonia. Consolidation.* They took X-rays using a portable machine that was shoved quickly through the door and back out again. Finally a doctor detached herself from the scrum. Bryce was very sick, she said. He would need to go to the intensive care unit, but first he needed surgery. Cell debris and fluids had filled up the air spaces of his lungs, and the slurry was as solid as gelatin, too thick to suck out through a tube. For Bryce to breathe, they would have to cut open his lungs and scrape out whatever they could.

The doctors advised sedating him and taking him to the OR immediately, and told Scott and Katie they could hold his hand, if they wanted. A lot of parents wanted to leave at that point because it was too hard to watch their child slip into unconsciousness. Scott and Katie locked hands and shook their heads—they would stay. Leaning over the gurney, they watched as the sedative slid into Bryce's IV line. His eyelids fluttered shut, his body relaxed, and the grunting faded to a whistle.

The gurney holding their son slid between swinging doors and vanished. Scott burst into tears.[1]

The next time Scott and Katie saw Bryce, it was early morning on New Year's Day, and he was lying flat in a curtained bay in the hospital's pediatric ICU. Twelve IV bags of drugs and fluids dangled above his head; they were plugged into infusion pumps, beeping, blinking boxes with small screens and keypads. Five drainage tubes pierced his chest. The respirator mouthpiece was taped to his face so it wouldn't slip, and his face was as pale as the tape. Though he was deeply sedated to keep him still, he was also secured between bolsters to ensure he couldn't thrash about and pull the lines loose.

The slender, energetic chief of pediatric infectious disease, Dr. John Bradley, was waiting for them with questions. Had Bryce ever been in the hospital? Had he had any skin infections or the flu? Had anyone else in the family been in the hospital? Had anyone recently gotten out of jail? Exhausted and frightened, they answered no to everything, though the questions didn't make sense to them. They could not make sense either of Bryce's chest X-ray, a black and gray image of his ribs and lungs mottled with bright patches of white.

Black meant healthy lung tissue, Bradley told them, and white was the infection that kept Bryce from breathing properly. The right lung was almost entirely white. When Katie squinted at the monitor, it looked to her as though the white continued beyond Bryce's lungs. Bradley told her that was right. The infection, whatever it was, had eaten a hole through Bryce's lung and into the space between his lungs and his chest.

It would take another day for test results to confirm it, but Bradley thought he knew what was making Bryce so sick. The lung inflammation was a clear sign of pneumonia, and Bradley thought the infection that was causing it was MRSA. Scott and Katie looked at him blankly—neither of them had ever heard of the bug. Bradley didn't tell them that the Rady team knew it very well. Not long before, they had treated a two-week-old who had lost all the tissue on his back to flesh-eating disease caused by MRSA, and a three-month-old who died of MRSA pneumonia and septic shock.

From those cases and others, the Rady physicians knew how to respond when they suspected an overwhelming MRSA infection: hit the organism with the strongest drugs possible, and give the hard work of breathing to machines so the body could heal. The emergency room doctors had given Bryce vancomycin immediately, and the respirator was dialed up as high as possible. But the glowing green numbers on the monitor over his head showed that his oxygen level was still perilously low. The gas was not penetrating the ravaged tissue. It was like pumping air into a brick.

Identifying MRSA and quelling it was the job of Bradley and his colleagues in the children's hospital's infectious-diseases department. Keeping Bryce alive and stable while the antibiotics did their work was the critical care team's challenge. The team head was a physician named Bradley Peterson, brilliant and highly credentialed, and also direct and so brusque that nurses were reserved around him and physicians like John Bradley contrived to be present at family chats to soften the force of his personality. When the Smiths arrived in the ICU, Peterson was already in Bryce's enclosure, giving orders to a nurse: change this, change that, order a transfusion. Something about the distraught young parents touched him. He stopped. Without introducing himself, he said, "I'm going to save your son."

* * *

It was a stroke of luck for Bryce that Peterson and Bradley's teams had seen that recent, deadly case of MRSA pneumonia. Few doctors had. Most of the medical community rarely encountered pneumonia caused by staph, so they did not recognize it or prescribe the correct drugs when it occurred.

Decades of experience and research had demonstrated that the main cause of everyday pneumonia—the kind that happens in a healthy person, not the kind that attacks an already-sick hospital patient—is a bacterium called *Streptococcus pneumoniae*, usually just called *Pneumococcus*. In contrast, staph—any staph, not just MRSA—had never been a major cause of pneumonia, and accounted for only about 2 percent of cases of the disease that started outside hospitals. When staph did cause pneumonia, the disease was almost always more severe than the more common pneumococcal pneumonia—people who developed it were likely to need ICU care, and were twice as likely to die—but it was such an unusual occurrence that it was far down on the list of possible causes that physicians considered. And if staph pneumonia was rare, MRSA pneumonia was barely thought of. In 2003, the Infectious Diseases Society of America issued official guidelines for what drugs to prescribe for everyday pneumonia.[2] The document listed the drugs that worked for illness caused by pneumococcus and drug-sensitive staph, but did not even address cases caused by MRSA.

But guidelines—consensus documents that are written by committees within professional societies on the basis of medical-journal reports—trail by years the real-world events they are based on. And in the real world, MRSA continued to change. In 2002, French scientists reported that two unrelated patients—a sixty-seven-year-old man who lived in Paris and a seventeen-year-old boy who lived 240 miles away in Nantes—died of what they called necrotizing pneumonia, which not only inflamed their lung tissue but killed and liquefied it.[3] In 2003, British physicians reported a similar case, in a thirty-year-old woman who came to a London hospital with chest pain and spent three weeks in their ICU.[4] Then Australia recorded its first case in 2004. A healthy twenty-one-year-old man came to the emergency room with fever and a cough, and got amoxicillin, a penicillin-based drug that is appropriate treatment for pneumococcal pneumonia. Two days later, he came back extremely ill and died of necrotizing pneumonia just two days after returning to the hospital.[5]

All of the victims were infected with staph that matched the resistance patterns of community-associated MRSA in the United States: they contained SCC*mec* type IV and manufactured the toxin PVL that was suspected of causing tissue destruction.

In 2005, U.S. doctors began to see similar cases. Over two months, physicians in Baltimore treated four people—three women, aged twenty, thirty-one, and thirty-three, and a fifty-two-year-old man—whose lungs had been eaten away and were studded with pockets of pus. The man died after two days. One of the women spent four weeks intubated in an ICU; a second was hospitalized for three months, and the third lost both her legs below the knee.[6] A few months later, doctors in Oakland, California, witnessed the same ferocious bacterial attack. A thirty-one-year-old man who had been feverish for two days and coughing for one day came to the ER, and was given the standard drugs for community pneumonia. He came back fifteen hours later unable to breathe, and died thirty-eight hours after that.[7] All of the patients were infected with community-associated MRSA: USA300, producing PVL.

Notably, the U.S. patients and those in England, Australia, and France were all adults. As of 2005, there was no evidence that this new, necrotizing pneumonia attacked children. But as Scott and Katie Smith were discovering, endlessly inventive staph had found its newest set of victims.

The Smiths channeled their desperate anxiety into a routine. They drove to the hospital together every morning, leaving home at 4:00 a.m. to beat the traffic and be caffeinated and alert when the ICU team did its morning rounds at 7:00 a.m. They stayed until evening rounds twelve hours later, and then stayed some more. They drove home at midnight, sleepwalked into the shower, collapsed for a few hours, and did it again the next day.

They spent the long days sitting by their son's bed, reading him stories and playing soothing instrumental CDs over the huff and hiss of the respirator. They watched the pulse-ox readout, trying to will the number to change. One hour it would flicker upward, toward the safe zone, but then it would drop, and their hearts would sink with it.

"The doctors would set a goal of a certain number, and we would walk in and the nurses would say, 'It's there, it's there!'" Katie recalled. "And they would turn his oxygen down a little. And then six hours, twenty-four hours later, they would be saying, 'No, it doesn't look so good.'"

Scott would probe for any good news. "You know, we love camping," he said to a nurse one day. "When he's all through with this, when he's okay, I'd like to take him to Yosemite. It's a four-thousand-foot elevation. Do you he'll be able to handle that?"

The nurse looked away. He thought she winced slightly. She said, "I wouldn't think that far ahead."

Because Bryce had been given the right drugs straight away, the evidence of MRSA in his blood began declining by his second week in the hospital. Yet his lungs were not healing. They were gravely inflamed from the toxins the staph had produced. His pulse-ox still hovered in the danger zone, and every repeat X-ray showed why: his lungs were solid white, with no sign of healthy translucent alveoli that would let the oxygen in. Because Bryce was lying motionless, fluid collected in his lungs, and the nurses had to turn him to dispel it and prevent additional bacteria from breeding. Every time they moved him, his lungs collapsed and his oxygen level dropped again.

On Bryce's twelfth day in the ICU, Peterson told the Smiths that it was time for more aggressive measures. He planned to bring in a high-frequency oscillator, a type of ventilator that would push breaths into Bryce's lungs four to five times every second. The ventilator would hold his lungs open like an inflated balloon, forcing air through the membranes without his chest having to contract. This would give the lung tissue time to heal without the stress of inflating and deflating again. Peterson warned them, however, that the machine would be noisy and upsetting. It was as loud as a jackhammer, and the vibrations would shake Bryce's body. More distressing, they would not be able to see their son breathing.

Bryce would only be able to stay on the oscillator for thirty-six hours, at most. After that, the next step was an ECMO, short for extracorporeal membrane oxygenation: a bypass that would shunt all his blood out of his body via catheters threaded near his heart, and aerate it through an artificial lung before pumping it back in. Though Peterson didn't say it, Scott and Katie understood: the ECMO was a treatment of last resort.

The ECMO, a hulking tangle of tanks and tubes, was fetched from storage and parked just outside the door. Even with her back turned Katie felt its oppressive heft. She passed it with dread. "My feeling in my heart," she said, "was if he went on the ECMO, he would never make it off."

After twenty-four hours on the oscillator, Bryce's lungs started to improve. The ECMO was taken away.

Still, Bryce's lungs were so destroyed that it took forty more days on a regular ventilator before the critical care team thought he might be able to breathe on his own. On a day in February, they decided to wake him. Doctors, nurses, and respiratory technicians crowded into the room. Scott sat on the bed, holding his son. Dr. Susan Duthie, a critical care specialist, leaned down and hugged him. There were tears in her eyes.

"There was a time we didn't think—" she said, stumbling over the words. Scott nodded. "I know, I know."

One of the physicians slid the breathing tube free. Bryce coughed. The doctors had warned that his throat would be too raw to talk. He looked up at his parents and slowly pulled the blanket toward his face. Katie caught her breath in a sob. He was playing peek-a-boo.

Bryce went home on February 24, 2006. He had been flat on his back for so long that his muscles had atrophied, and he had to learn to walk again. He had also forgotten how to swallow. For a month he took a daily dose of methadone, to blunt his withdrawal from the morphine that had kept him sedated so long. He needed speech therapy, and he had lost the hearing in one ear.

In September 2007, Scott and Katie took him to Yosemite.

The possibility that pneumonia could be caused by MRSA, rather than pneumococcus or even ordinary staph, was a troubling development, unnerving to epidemiologists and devastating to its victims. It was hard to believe that MRSA could find another mode of assault. But in all the millennia of its existence, staph had never become less serious than it had been before—and shortly after it began to cause MRSA pneumonia, the bacterium found a partner to potentiate its lethal attack.

It teamed up with flu.

Flu and staph are old companions. Every few decades, flu viruses mutate enough to appear completely unfamiliar to the human immune system, causing a global epidemic known as a pandemic. The H1N1 influenza pandemic that began in Mexico in April 2009 was only the latest, and in its early days the mildest, in a pattern that has been repeating for centuries. For every pandemic for which there are good records, there are accounts of huge numbers of pneumonia deaths.[8] In the midst of the 1918 pandemic, which killed an estimated 50 million people around the world, a physician working at the Camp Devens military reservation outside Boston wrote to a friend:

These men start with what appears to be an ordinary attack of LaGrippe or Influenza, and when brought to the Hosp. they very rapidly develop the most viscious type of Pneumonia that has ever been seen. . . . It is only a matter of a few hours then until death comes, and it is simply a struggle for air until they suffocate.[9]

Physicians reported the same kinds of illnesses after the 1957 pandemic, which killed a presumed 2 million people, and also after the less deadly 1968 pandemic, which killed 1 million around the world.[10,11] As far as physicians and researchers could determine from records and crude microbiology, bacterial pneumonia caused by staph and pneumococcus was a major killer. (Early analyses of the 2009 H1N1 pandemic found the same.)[12] Then in 1997, enterprising researchers retrieved rare preserved samples of lung tissue from soldiers who died in the 1918 pandemic, which had languished for decades in military autopsy archives.[13] First they isolated and reassembled the flu virus the tissues contained. Then they assessed the condition of the lung tissue itself. They confirmed what earlier pathologists had not had the technology to prove: a large proportion of deaths in the 1918 pandemic—which killed up to 100 million—were due to severe pneumonias caused by common bacteria including staph.[14] The bacteria normally lived elsewhere on the body, or high in the nose and throat, but were able to slip deep into the lungs because flu infection had destroyed the protective mechanisms that kept pathogens at bay.[15]

Researchers presumed that the 1918 flu virus caused so many millions of flu cases, and therefore so many millions of pneumonia deaths, because it was a novel virus against which the immune system had no defenses. In ordinary flu seasons, the virus mutates or "drifts" only slightly from the previous year. Bolstered by a flu shot, the immune system recognizes the virus and mounts a defense against it, making flu cases less likely and reducing the likelihood that deadly bacterial pneumonias will occur. In an ordinary flu season, bacterial pneumonias—particularly those caused by staph—were rarely expected.

But it was an ordinary flu season in the fall of 2003 when healthy children in Colorado began dying.

The flu struck unusually early that year. To make matters worse, expert predictions made in the spring, about which flu viruses would circulate during the 2003–2004 season, had been incorrect—so the flu vaccine formulated according to those predictions did not protect as completely as it should have. Two days before Thanksgiving, the health department for the city and county of Denver held a press conference. Four children—a fifteen-year-old, an eight-year-old, a two-year-old, and a twenty-one-month-old baby—had died of flu-related pneumonia within the past week, and the Denver Children's Hospital was overloaded with seriously ill kids.[16] Other states soon began reporting the same: ER waiting rooms were standing room only, and pediatric ICUs were running out of ventilators.[17]

Every year, some children die of flu, usually because they have underlying illnesses such as diabetes or asthma that make them vulnerable. These deaths, though, were different. In December 2003, the CDC reported "severe complications of influenza" in healthy children around the country: seizures, swelling of the brain, crashing blood pressure, and bacterial pneumonia. Three children—a sixteen-year-old, a twenty-two-month-old, and a twenty-month-old—had died of pneumonias specifically caused by MRSA, something the CDC had never seen before in flu.[18]

At the time, doctors were not required to report either children's flu deaths or MRSA pneumonia to the authorities. After the 2003–4 flu season ended, the CDC struggled to understand what had happened, asking health departments and hospitals to pore over their data. Forty states responded. Altogether, there had been 153 deaths; epidemiologists thought that seemed high, though they had no data to prove it.[19] Out of the forty, only nine provided information about which bacteria caused the fatal pneumonias. Those states found fifteen cases caused by MRSA. All cases involved the community strain, either USA300 or USA400, and all the strains made PVL.[20]

The CDC could not tell how many MRSA pneumonia deaths occurred in the thirty-one states that reported fatalities but did not do detailed microbiology, or in the ten states that did not send reports at all. Though there was no way of proving it, it seemed likely that the fifteen proven MRSA deaths were a signal of a dangerous new phenomenon. "This pneumonia is very rapid—from onset of symptoms to death, within a week," said CDC epidemiologist Jeffrey Hageman, who wrote an analysis of the deaths. "There is nothing that could have been done to change their course."[21]

It was potentially dangerous for several different reasons. As the French and British cases in adults had shown, MRSA pneumonia was extraordinarily destructive to lung tissue, making it uniquely lethal. But in addition, MRSA pneumonia was still such a new thing that doctors were unlikely to suspect the bacterium's involvement, meaning that those who did not think to test for it were unlikely to prescribe the best drugs. And finally, MRSA's growing arsenal of resistance factors meant that, once necrotizing-pneumonia patients did reach the ICU, there would be fewer drugs that could be used to quell the bacterial assault.

It was a mystery why MRSA pneumonia had arisen at just this point. A few researchers had a hypothesis that would explain it, but they hesitated to voice it, because it would have thrown doubt on what looked like a public-health success. In 2000, the drug manufacturer Wyeth Lederle had introduced Prevnar, the first vaccine against seven separate strains of pneumococcus. It was spectacularly successful, cutting rates of pneumococcal pneumonia by two-thirds.[22] But there was no way to know whether that reduction in pneumonia was permanent—or whether preventing pneumococcal infections had opened an ecological niche that some other organism, perhaps MRSA, could occupy instead. If that hypothesis was right, cases of MRSA pneumonia would start to climb.

Only one child died of flu-related staph pneumonia in the 2004–5 season, and three in 2005–6. But during the next season—the winter of 2006 and spring of 2007—states began besieging the CDC with reports of severe pneumonias in children that were demonstrably caused by staph. In nineteen states there were thirty-seven cases of severe MRSA pneumonia among kids with flu, and sixteen deaths.[23]

In Louisiana, where the health department was closely watching for MRSA, a ten-year-old boy came down with a fever, cough, sore throat, and earache. His family gave him Tylenol, but his symptoms worsened overnight and they took him to the emergency room. He was immediately admitted to the hospital and sent to an intensive care unit, where he died two days later. In Georgia, a family brought an eight-year-old girl to the doctor after three days of fever and cough. After her heart stopped and was restarted and her liver and kidneys failed, she was put on an ECMO. She died in the ICU twenty-five days after she was admitted.[24]

The deaths left the CDC's Hageman, the father of a toddler, confounded by how they might have been prevented. Some of the children, or their relatives, had had MRSA skin infections earlier in the year. That might have served as a warning; the possibility that a staph skin infection could set a child up for staph pneumonia had been well known in the penicillin era.[25] But in the ensuing decades, that link had passed out of medical knowledge, and most family physicians still knew MRSA only as a hospital danger, not a community one. The children's treatment testified to that—most of them had been given drugs that would work against everyday pneumococcal pneumonia, but not MRSA. But to be fair, there was no obvious sign that could have warned their doctors otherwise. The sick children's initial X-rays had looked like common pneumonia, and many of them had been infected for such a short period of time that tests could not detect MRSA in their blood.[26]

There was one possible preventative, but Hageman and his colleagues knew in advance that it would be a tough sell. Flu vaccine might protect children against MRSA pneumonia, because it would keep them from contracting influenza and experiencing the tissue inflammation that allowed MRSA to take hold in the lung. None of the children who died in those nineteen states in 2006–7 had received a flu shot. But that was not unusual; most young children do not. Though the vaccine had always been available for children, federal campaigns urging people to be vaccinated had focused for years on the elderly and chronically ill who statistically were more likely to die of flu. In 2007, as part of a campaign to expand flu vaccination, federal health authorities for the first time recommended that every child younger than five be given flu vaccine, largely out of hope that vaccinated children would not pass flu infection to their older relatives.[27] But for many reasons—the difficulty of organizing yet another pediatrician's visit for kids already getting many childhood vaccines, the lack of a clear message that children were themselves at risk—the recommendation was not effective. Only one child in six actually received the shot.[28] And the few children who did receive the vaccine were school-aged. Though the vaccine had always been available to adolescents, there was no official recommendation that healthy teenagers be given the flu shot. But adolescents, like younger children, were being felled by MRSA pneumonia that they contracted after a bout of flu.

The desert hills north and east of San Diego are dry, studded with boulders, and treeless except for low tufts of scrub. In summer they are broiling hot and a severe fire risk—so for decades, San Diego schools have taken their sixth-graders on camping trips in the winter. Amber Don and her husband Carlos Don III had both taken such trips when they were growing up near the ocean between the North Island Naval Air Station and the Miramar Marine Corps base. Now they lived in Ramona, a small horse-raising town on the way to the Salton Sea, and it was time for their son, Carlos Don IV, to take the gleeful parents-free outing that would launch his last semester in elementary school.[29]

On January 16, 2007, they put twelve-year-old Carlos on a school bus headed for the mountains. Carlos, the couple's oldest child and only son, was born when Amber was just eighteen, and she thought of him as her buddy. He helped her around the house, watched his sisters, and held her hand when no one else was looking. But he was his father's son as well, MVP on his Pop Warner football team and a champion motocross rider since he was five.

With more than two hundred sixth-graders going on the trip, his parents expected Carlos to race around all day, stay up talking all night, and come home blissfully exhausted. But when his father picked him up at the bus stop four days later, he immediately called his wife at her mortgage-brokerage office forty miles away.

"Amber," he said, "he looks like a ghost."

Carlos had a 104-degree fever. He'd been sick since Wednesday night, but hadn't wanted to spoil the trip by telling anyone. Amber was exasperated with him for keeping quiet, and frustrated that no one noticed how ill he was. Thinking he had the flu, she dosed him with Motrin, tucked him in on the couch, and gave him a bottle of Gatorade. The next morning, the fever had gone down a bit, but he was coughing and shaking with chills. He was so weak Amber had to carry him to the bathroom. She decided to take him to urgent care.

The staff there thought Carlos had walking pneumonia. They showed her an X-ray that was lightly cloudy, and gave her a prescription for codeine to suppress his cough, and azithromycin, a standard antibiotic for everyday pneumonia.

Amber took Carlos home and tucked him back on the couch in front of the big TV. He stayed there all night, coughing and shifting around to get comfortable. Amber slept in the living room as well, in case he needed anything and couldn't get up. The next morning, the coughing had stopped, but he complained that his chest hurt. He was still weak, and though his fever had gone down a bit, he was clammy and his skin was cold. Amber started doing chores in the kitchen to keep him company, and left the house to clear some trash out of her car. When she came back in, Carlos was still on the couch.

"I was hungry," he told her. "I ate a hot dog."

Amber started to say that was good—he hadn't eaten solid food in days—but then she halted, and a chill washed over her. There were no hot dogs in the house; she hadn't bought any, because he was going to camp. She checked the kitchen. The counters she had wiped down earlier were still pristine. Carlos was hallucinating.

She dumped him in the car and revved down their twisting mountain road, powering onto Ramona's short main street. At the same urgent care office, the nurse clipped a pulse-ox light on Carlos, checked his vitals, and quickly reached for the phone.

"I'm calling an ambulance," she told Amber. "His blood-oxygen is only 60 percent."

The closest hospital was twenty-two miles away on two-lane roads, in Escondido. When the ambulance arrived, the staff put Carlos on oxygen and gave him a round of asthma drugs. But in the ER doctor's opinion, a boy that young and that sick needed to be in a children's hospital. They loaded him back into the ambulance again and sent them to Rady Children's Hospital, another thirty miles. Amber sat in the front, leaving the seats next to Carlos's gurney for the paramedics. As they reached the San Diego suburbs, she heard a commotion and turned around to see one of them on top of Carlos. "Lights and sirens," the other one told the driver. "We have to get there right away."

At Rady, they rushed Carlos straight to the ICU. He was conscious and Amber could tell he was frightened and trying not to show it. She kissed his forehead and held his hand.

"Okay, dude," she said, trying to sound casual. "Dad will be here in a little while. I love you, you know that?"

"I love you, Mom," Carlos whispered back.

It was the last thing she ever heard him say.

Since the ICU enclosure was small, the nurses asked her to go to the waiting room. They told her it would be about forty-five minutes. An hour passed. Then, two hours. After three hours, a resident came to show her Carlos's chest X-ray on a computer monitor. The day before, the X-ray at urgent care had been threaded with faint strands of grey. Now it was completely white.

"Is he going to be okay?" Amber asked, fighting to control her voice.

The resident shook his head. "To be honest," he said, "I doubt he will make it through the night."

In the year since the critical care team at Rady treated fourteen-month-old Bryce Smith, they had not seen a child that sick; but twelve-year-old Carlos was sicker. He was in what doctors call acute respiratory distress syndrome, or ARDS. His lungs were inflamed and stiff, awash in fluids leaking from his blood vessels. The ICU staff sedated him into a medical coma and put him immediately onto a ventilator, dialed to a special setting to infuse oxygen into his blood without expanding his lungs and injuring them. They added nitric oxide and helium to coax the air through the tissue, and threaded a catheter through his heart to monitor the blood going into his lungs. Nothing worked. In less than twelve hours, his lungs became useless, and they put him on the ECMO, the heart-lung bypass machine that Bryce had narrowly avoided.

Dr. John Bradley, the chief of pediatric infectious diseases who had recognized MRSA in Bryce, was on his way into work when his pager buzzed. The admitting physician had pored over Carlos's recent past, looking for something that could make a teenager's lungs fail so quickly and completely, and something had struck him. When the boy felt the first symptoms, he had been up in the mountains. That meant he could have picked up an illness from the environment, or from a bird or a rodent—possibly valley fever, hantavirus, avian influenza, or even pneumonic plague.

When he heard the doctor's suspicions, Bradley went directly to the ICU, skipping his office and his usual morning rounds. Most of those diseases were potentially fatal, and treating hantavirus required a rare drug formulation that Rady did not stock. Several were public health emergencies as well. Plague is on the short list of organisms that can be used for bioterrorism, and as a major trade port with Asia, San Diego had been watching for avian flu for years. Invoking public health concerns meant the doctors could get access to the bioterrorism rapid-detection apparatus that states had built up since the anthrax attacks in 2001. Bradley sent samples of Carlos's blood to the San Diego public health laboratory. A rapid test for flu was negative, but just in case he put the boy on oseltamivir, called Tamiflu, an antiviral that can mitigate severe flu if started within forty-eight hours of infection; it was a slim chance given how long Carlos had been sick, but worth a try. Then Bradley ordered every drug that might possibly be relevant, more antibiotics than he had ever before given to a kid with pneumonia.

Before night fell on Carlos's first day in the hospital, the health department reported back. All tests were negative. Bradley headed to the hospital's microbiology lab. It was barely twenty-four hours since swabs had been taken at Carlos's admission, not enough time for a culture to mature, but mentally crossing his fingers, he pulled a plate from the incubator. It was overgrown with staph. The lab also did a MRSA rapid test and confirmed it: Carlos had the bug.

Bradley ran back to the main building to deliver the news and make sure that Carlos was getting the highest dose of vancomycin, as well as clindamycin to shut down staph's destructive toxins. Outside the ICU, he collided with a crowd: the Dons' parents and relatives and Carlos' team-mates and school friends, with food and blankets and plans to spell his parents and take care of his sisters. The families were making a schedule. Amber's father would take the earliest shift, three hours starting at 4:00 a.m., so she and her husband could rest before he went to work. Her husband's father would take the late nights, from 8:00 p.m. til 11:00, so she could shower and eat while her husband took care of the girls.

The next morning, the health department called back. They had done an additional round of tests, using more sensitive equipment that Rady did not have. Carlos did have the flu.

Amber refused to leave the hospital. She hated even to leave the wait-ing room, in case they needed her suddenly and she could not be found. Every day, all day and most of the night, she sat by Carlos's bed or curled up in the room's small window seat, talking to him, reading to him, and concentrating as though she could make him better by force of will.

Briefly, it seemed to work. In the initial onslaught, Carlos's blood pres-sure had crashed and his kidneys had shut down, but being on the heart-lung bypass allowed the critical-care team to regulate his blood pressure mechanically. ECMO was far from benign—to pump the blood safely through it required heavy doses of anticlotting medications, so there was always a risk of causing a bleeding stroke. But statistically, most patients who went on it came off it alive. It was also possible to plug a dialysis machine into the ECMO, to continuously clean the blood; so when Car-los's kidneys started to fail again, risking a toxin buildup that would dis-rupt the fragile balance the team had achieved, the bypass machine kept him stable. Toward the end of the first week, the ICU staff began to speak hopefully of taking him off the ECMO as soon as his lungs began to heal.

But in that week, his chest X-rays had not changed. His lungs were so deeply damaged by staph that they showed no working air space at all. And by the end of the week, the secondary effects of MRSA's continued assault began to show. The toxins produced by the bug had damaged his heart and blood vessels, and the small veins and arteries in his arms and legs began to clot off. Marks like bruising broke out on his skin. They spread and merged and darkened; in between, the blood-starved flesh showed unhealthy white. Cut off from his circulation, Carlos's legs and arms turned black.

"His parents were wonderful," Bradley said. "They said, 'We'll take him home with no arms, no legs. It doesn't matter. Just please save our child.'"

Since admission to the ICU, Carlos had had a chest tube, a wide catheter to drain pus and blood from his lungs. In his second week in the hospital, his lung tissue began to leak out the tube. The staff put in a second tube. It too filled with dead, dissolved lung. In a desperate reach, Peterson considered putting Carlos on the waiting list for a lung transplant, only to realize that his damaged circulatory system could not sustain the new organs.

At the end of his second week, with the chest X-rays still unchanged, the ICU team sent Carlos for a CT scan. It showed what they feared: his lung tissue was gone. Worse, the clotting that had choked off blood to his extremities had killed his large intestine as well. His heart could not pump unaided. His arms and legs were dead.

"They told us, 'We'll keep him on life-support as long as you want,'" Amber said. "But I knew what we had to do."

On Saturday, February 3, they summoned their relatives, and Carlos's football coach and friends and teammates. The ICU staff let the visitors go in three at a time to say goodbye. There were so many that the visits took all day, and the ICU's two waiting rooms filled with knots of sobbing teenagers. The Dons asked their parish priest to give Carlos last rites and perform his confirmation, the Catholic sacrament of adulthood that he would have received in the spring. As she listened to the prayers, Amber thought she saw a tear slide down Carlos's cheek.

At 2:20 a.m. on February 4, with his mother and father holding him, the ICU staff unhooked Carlos from the dozen IVs that had sustained him for two weeks, the huge catheters that carried his blood to and from the bypass, and finally the breathing tube that had kept his lungs inflated in hopes they would heal. They closed the door behind them. Carlos gasped three times, and died.

CHAPTER 8

THE FOE ON THE FIELD

The grave fates of Bryce Smith and Carlos Don were extraordinary, but the MRSA that started their illnesses was becoming strikingly common among healthy young children. It was dawning on epidemiologists that kids were most vulnerable in places where they spent large amounts of time with other children. For young kids, there was only one candidate: school. And as researchers were soon to discover, there were particular risks associated with particular locations in schools, the locker room and the athletic field.

If you don't already live there, there is no particular reason to go to Stafford, Texas. The little city on the southwest edge of Houston's massive sprawl isn't exactly on the way to anywhere. The highway to San Antonio lies to the west, the road to the beaches of Galveston is east, and Houston's outermost beltway—the most recent concrete corral around its constant expansion—was laid down three miles north. Its citizens like this fine, considering Stafford something of a shared secret: it is rich enough to have voted out property tax in 1995 and quiet enough to have had the same mayor since 1969.

The main draw for outsiders is its churches. Stafford is not more overtly religious than the rest of Texas, a place where sizes of everything—cars, hair, and helpings of barbecued brisket—start at XL and go up. Yet thanks to loose zoning laws, Stafford has fifty-five houses of worship, one for every three hundred residents: indie storefronts, mainstream Protestant, Chinese Baptist, Taiwanese Buddhist, multiple sects of Roman and Orthodox Catholic, and a turreted Hindu temple that was hard-carved in India and assembled piece by piece on a twenty-two-acre lot. The various denominations demonstrate greater Houston's peak-oil diversity as surely as the sign outside Nick's Gas Station near the center of town: "Barbacoa, menudo, gorditas, hamburgers, fried chicken, fried rice."

Stafford is the only city in Texas with its own school district, and it groups its few schools—one primary, one elementary, one intermediate, one middle, and one high school—on one broad campus dotted with knobbly live oaks. The buildings are twenty-five years old, but with only three thousand students cycling through them every year, they still look fresh: dusky blond stone topped with horizontal stripes of glazed red brick, and ice-white corridors lined with scarlet doors and yellow lockers. Red and yellow are the colors of the school's Stafford Spartans, and they are inlaid on the running track, streaked on the cafeteria walls, and woven into the collars of the school staff's polo shirts. The athletic fields are real grass. The training room has as much equipment as a small commercial gym. The campus may be wedged between St. Thomas of Kerala Indian Orthodox Church and St. Joseph Syro-Malabar Catholic Church, but inside its fence, it shares the true religion of every school in Texas: football.

Every boy at Stafford Middle School and High School aspires to play football—the Anglos who grew up with it, the Latinos who sneak in soccer-quick footwork, the newly arrived Nigerians and Russians, and the Tongans whose size makes coaches widen their eyes and recalculate plays in their heads. Since the schools are so small, it is hard to keep a deep bench, and talent is not constant but comes in waves every few years. The teams do not always play well, but they always play hard.

So it was unsurprising when Nick Johnson, a twelve-year-old seventh grader, wandered up to the sidelines during practice in October 2003 to say he had tackled a classmate while trying out a play, and hurt his left shoulder. David Edell, the school's licensed athletic trainer, checked Nick for bruises and broken skin, found nothing, and told him to ice it down and discuss it with his coaches the following day.

Instead of practice, the next day Nick went to the pediatrician for X-rays. The films would tell his doctor and his mother nothing, though, of what was happening in his sturdy, undersized body, and nothing about how his life, and his school's devotion to sports, were about to change.[1]

Nick's shoulder ached, and he couldn't move it much, but his pediatrician could see nothing wrong on the X-rays—no dislocation, no breaks or chips—so he gave the boy Tylenol with codeine and told him to keep the arm in a sling. A day later, his chest hurt as well and he felt hot. He went to bed early, but then at about 11:00 p.m. he woke his mother Janet. She was shocked by his fever. It was over 104. It was a Friday night, a bad night to go to the ER, but she decided to take him to the hospital in downtown Houston where she worked in medical-records IT. It happened to be Texas Children's Hospital, one of the few in the United States where infectious-disease researchers had been unwinding the unfolding threat of community-associated MRSA in children.

Nick was so uncomfortable in the waiting room that he refused to sit in a chair; while they waited to be seen, he lay on the floor at Janet's feet, shifting restlessly. A second set of X-rays showed slight shadows in his lungs, but his heart and the lymph nodes in his chest were normal. The physicians diagnosed a minor case of pneumonia, gave Janet a scrip for a beta-lactam antibiotic, and sent them home early Saturday morning.

Nick started the drugs immediately, but they gave him no relief. He begged his mother to run hot baths for him. Janet was reluctant—she was afraid the hot water would drive his fever higher—but he said it made him feel better, so she gave in. While he soaked, she fretted, wondering when the antibiotics would begin to work. She offered him food, and a cold soda. He wasn't hungry. He didn't want anything to drink.

All weekend Nick lay limp on the couch. He missed school on Monday, too, complaining that his knees and elbows hurt. When his father Dale came home at lunch to check on him, Nick's left arm was brilliant red and swollen from shoulder to wrist. Dale ran him back to the hospital. Janet rushed down to the ER from her office, and some of her work friends joined her. One was a pulmonologist; pneumonia was his forte. "Nick doesn't look good," he told Janet gravely. "You need to get him seen right away."

Nick was hunched over in the ER wheelchair, folded around the hot heaviness in his chest, half-hearing the commotion around him through a haze of fever and pain. Needles scared him—he hated getting shots for school—but now he barely noticed as nurses threaded IVs in the tops of his feet and in the backs of both his hands. His sight was blurry, and he couldn't feel the needles going in. One nurse missed a vein, and he glimpsed the back of his left hand covered in bright blood. He passed out.

Nick's blood pressure and blood oxygen content were falling. An emergency blood test showed that his blood chemistry was out of balance. Almost half of his white blood cells, an incredibly high number,

THE FOE ON THE FIELD

were bands, immature cells that the bone marrow pumps out when an overwhelming bacterial infection taxes the immune system. On an X-ray, the radiologists could see collections of fluid in his shoulders and elbows, and septic emboli—clots of infected tissue that had broken off from somewhere else in his body—that had lodged in his lungs. There was a bright area of infection in the bone of the shoulder he had banged in the tackle the week before.

The infection was MRSA. Nick was immediately intubated, and the doctors told the Johnsons to call their daughter, at college in Kansas, and get her home fast.[2]

For about a year, coaches in schools across Texas had been wondering whether there was some connection between MRSA and football. During the 2002 football season, twenty-nine players at Pasadena's Sam Rayburn High School, thirty miles east of Stafford, were benched by drug-resistant infections. So were six players at the University of Texas in Austin, 180 miles west, and two players at Smithson Valley High School in Spring Branch, forty miles beyond Austin. Between the start of preseason camps in August and playoffs in December, twenty high school teams called the Texas Department of State Health Services to ask what was going on.[3,4]

A wrestling team in southwest Vermont could have answered their questions. In the summer of 1993, a seventeen-year-old high school varsity wrestler was treated at a local hospital for an abscess on his forearm. It healed without any problems—but once school practices started in the fall, seventeen other wrestlers developed skin infections. So did six other people in the small town, including two other high school students and the first case's twenty-one-year-old brother. During the season, three teams from Vermont and eight teams from other states complained that their own students also developed skin infections following bouts in which they wrestled the team.

Investigators from the CDC and the Vermont Department of Health who analyzed the outbreak were mystified. They assumed the source of the infection was a hospital leak, because in 1993, hospital MRSA was the only recognized variety of the bug. But there was no MRSA outbreak in the local hospital, and the strain's resistance pattern did not match hospital MRSA. It was impervious to penicillin and beta-lactams, but sensitive to a range of other drugs that hospital staph was resistant to, including clindamycin. It was the same drug-sensitivity pattern that

113

would lead Robert Daum to realize that serious illnesses among Chicago children were being caused by a new community-associated strain—and that would eventually convince doubtful editors at the *Journal of the American Medical Association* to publish his alarm-bell paper.[5]

When she looked back fifteen years later, Dr. Joann Lindenmayer—a CDC field investigator assigned to Vermont in 1993 and a professor of environmental health at Tufts University in 2008—realized that they were seeing in Vermont what Daum saw in Chicago: the first random uprisings of the community strain that would sweep the country.[6] But the Vermont report, which was published only two months after Daum's seminal paper but in a different journal, the *Archives of Internal Medicine*, got as little attention as his did. Which was unfortunate, because in it Lindenmayer sounded a warning: "It is tempting to speculate that . . . shared towels and locker room facilities might have contributed to MRSA transmission."

By the time Nick fell ill in fall 2003, a few teams and trainers had begun to recognize that MRSA had a unique affinity for athletes. There were outbreaks among the Miami Dolphins, at Thiel College in Pennsylvania, at the high school in Franklin, Wisconsin, among varsity wrestlers in Indiana, and in an amateur fencing club in Pennsylvania.[7] Every case spurred additional anxiety among the few experts who noticed the disturbing trend. Athletes always compete against someone—so every game carried a risk of passing MRSA from infected or colonized players to an uninfected team, school, or town.

For most of the teams where athletes developed infections, MRSA was an unfamiliar opponent, but Dave Edell, Stafford high school's athletic trainer, already knew about the bug. In addition to working at Stafford, he was director of athletic services for West Houston Medical Center, which meant he rotated through training positions at a dozen rural high schools that could not afford full-time athletic trainers of their own. Early in the 2003 season, two players from those schools came to Edell with skin infections. Rumors of the infections percolated through local communities, and a TV reporter from Houston, where high school football is always big news, called Edell in mid-October to see what the trainer knew. The reporter was curious whether the serious infections that were afflicting the Texas college teams were showing up in Houston schools as well.

Oh no, Edell said. Sure, he'd seen skin infections. But none of the schools he worked with had serious illnesses among their kids.

The next morning, he got a phone call. Nick Johnson was in the ICU.

114

* * *

Nick had pneumonia in both lungs. Tissue in his lower left lung was dying, and septic emboli were wearing holes in his lungs and collecting in his kidneys. The bone in his left shoulder was infected, and there were collections of pus in his right shoulder and left hip. Over more than a week, while he drifted in a drug-induced coma, his physicians used ultrasound-guided needles to remove pus and fluid from both shoulders, the bone of his left upper arm, his right elbow and his left hip. They cut open his left shoulder and his left elbow to drain infection from the joints. They punched holes in his chest wall and plugged drainage tubes into the space around his lungs. Fluid collected around his heart, squeezing it so it could not fill completely with blood, so the surgeons opened his chest down the breastbone, cut away the protective sac around his heart, and siphoned out 20 ounces of fluid.

On his twelfth day in the hospital, they woke him up.

"You know those old-school water bottles with the ridges on the straw, and you pull on it and you can hear it come out of the hole in the lid? That's what the tube felt like coming out of my throat," Nick said.

It took several more days before the nurses could take out the tubes draining his chest, and when they pulled those free they hurt more than the ventilator did. On the seventeenth day, they moved Nick out of the ICU to an intermediate-level unit and took out his feeding tube. Suddenly, he was ravenous. He wanted a cupcake, he told his mother, and a sandwich, and some hot wings, and a taco. He ate half the cupcake, and fell asleep. When he woke up, it took him two hours to get the taco down.

He was not yet recovered. He had a cough, and during an MRI the radiologists could see bright spots of infection in both shin bones, both thigh bones and the right side of his pelvis. His fever finally broke on day 22. In mid-November, after thirty-seven days in the hospital, he went home. His voice was scratchy and strange from the tubes that had been in his throat, and the skin on one foot was burned and dark from a caustic medication that leaked when an IV needle had slid free. He had lost the hearing in his left ear. He was scarred across his left shoulder and down his chest, long slashes outlined in dots from stitches and staples, and round depressions where the drainage tubes had pierced between his ribs. He would have to take antibiotics for two more years.

Nick visited his school just before Christmas. None of his friends had seen him while he was in the hospital; you had to be fifteen to visit a patient in the Texas Children's ICU, and like him they were twelve or thirteen. As he took a slow walk down the school's corridor, the bell rang, and kids streamed out of classrooms for lunch. A couple of his friends ran by without recognizing him. He weighed thirty pounds less than when he had fallen ill.

He went back to school full time the next semester, still underweight and shaky. But he began lifting weights, and started his growth spurt. In the fall of eighth grade, a year after the fateful tackle, he played football again.

From the time Nick got sick in 2003, schools of all sizes and income levels began detecting much more MRSA in their athletes. In Texas, so many teams were talking about it that the state health department did a survey, polling the largest high schools in the state—the 4As and 5As, with enrollments of 900 to 5,200 students. The inquiry had to be approached with care. In Texas, high school football is not only religion, it is also an enormous business. More than 22,000 jobs depend on it, and coaches can make $100,000 a year, one-third more than the state's best-paid teachers. Some stadiums seat more than 20,000 people, and schools sell stadium-naming rights for millions of dollars. Local TV stations run regular high school segments, and games are broadcast nationally on ESPN.[8-12]

Only two-fifths of the schools that were sent the survey responded, but what they had to say was alarming. Thirty percent of them had cases of MRSA in their athletic programs, and not just in football, where it was most common, but in low-contact sports such as volleyball and cross-country running. Twice as many had cases of staph that were not cultured to identify the strain—but when state investigators wrote up their results, they made it clear they believed some of those cases were MRSA as well. "MRSA in athletic departments is much more pervasive than outbreak reports might indicate," they warned.[13]

Schools outside Texas also were finding the bug pervasive. At Sacred Heart University in Fairfield, Connecticut, ten of the one-hundred-person football squad developed boils, all of them caused by MRSA USA300, the dominant community strain. All the players needed at least a week of careful wound care before they were well enough to play. Two were sick enough to be hospitalized, one with the same septic arthritis that had invaded Nick's joints and bones.[14]

Epidemiologists from the Connecticut Department of Public Health picked apart the outbreak, trying to sort out why those ten players were vulnerable. They studied whether the players had been hospitalized recently, whether they had ever had skin infections before, what their personal hygiene was like, and what they did on the field. When the investigators considered all the data, a few things jumped out at them. Eight of the ten infected players were either cornerbacks or wide receivers, positions where they were more likely to have skin-to-skin contact with other players. Cornerbacks in particular were seventeen times more likely to have lesions than their teammates. The team played on artificial turf, and players with infections were more likely to have gone down hard and gotten turf burns on their arms and shins. The ten players had something else in common, too: they were much more likely than their teammates to shave their body hair—for hygiene, for fun, or because having smooth skin made it easier to tape up, and much less painful to take tape off.

The investigators realized that all three factors—positions, turf, razors—really involved the same thing, a micro-trauma to the skin that opened a portal for MRSA to slip through if it was already on the player's skin, or on the skin of any other player he crashed into. They found another element as well: no soap in the locker room showers. The epidemiologists had the school install soap dispensers, and the outbreak ended. When the dispensers ran dry and the custodians were slow in refilling them, the infections returned.

The St. Louis Rams experienced a similar outbreak, also in the 2003 season. Between August and November, five of the fifty-eight players— four linemen and a linebacker—developed hot, painful abscesses, two to three inches across, on their forearms and knees. None of them were hospitalized, but three of the five were sick enough to miss either games or practices. One was out for twelve days. Like the Connecticut collegiate team, the infected pro players were much more likely than their backfield teammates to be crashing into other players, and they were also much more likely to have turf burns. The CDC, though, couldn't find any evidence that the turf itself was harboring MRSA. Once again, epidemiologists ended the outbreak with simple solutions—keeping the players' wounds covered and their hands clean.[15]

But staying clean and keeping on top of wounds turn out to be more challenging to do, and to keep doing repeatedly and faithfully, than they might sound. The University of Southern California Trojans could testify to that. The team experienced their first cases of MRSA in 2002. It took them three years to completely eliminate the bug.

Like most college teams, the Trojans' season begins with training camp in early August. Players practice outdoors twice a day at Los Angeles's Howard Jones Field. They live together in four-person suites, and hang out together the rest of the time. Humidity, high temperatures, sweat, and direct contact made an ideal setting for breeding and passing around a bug that lives on skin. Five weeks after camp started, and three weeks after games began, one player complained of a painful abscess on his lower leg. A second showed up a few days later with a low fever and a small pimple on his elbow that hurt out of all proportion to its size. Both were hospitalized. The leg abscess needed surgical drainage and IV antibiotics. The pimple was the first sign of flesh-eating necrotizing fasciitis; that player needed multiple surgeries and skin grafts before his infection healed.

The coaching staff installed antibacterial soap in the showers and bathrooms, barred players with open wounds from using the whirlpools, and stopped using ultrasound gel and massage lotions in big pump bottles in case sharing contaminated them. The team suffered no more infections that fall. But in 2003, problems began erupting even before training camp ended, starting with an infection of a team member whose locker was directly across from one of the sick players from the previous year. Three weeks after games began, thirteen of the 107 players had deep skin infections, and five were sick enough to be hospitalized. The team swabbed the noses of every player and every staff member, and checked the athletic training facility and locker room to see whether any surfaces harbored the organism. The staff, student interns, rooms, and equipment were clear.

The Trojans asked the disease detectives of the Los Angeles County Department of Health Services to investigate. In the preceding few years, Health Services had become quite familiar with MRSA. They had found it in the Los Angeles County jail, in a newborn nursery at a local hospital, among soccer players, and in a group of gay men. They questioned the team, comparing the characteristics of the infected and colonized players with randomly selected players who had no infections and whose nose cultures were clear. As with other teams, the affected players were much more likely to be linemen, linebackers, or cornerbacks. They had lesions on their elbows, arms, and legs, the areas of skin that were most likely to scrape against another player's skin in a scrimmage. They were more likely to have lockers close to other infected players, and to share towels and bars of soap, and sometimes they napped in the training room on piles of used towels.

The descriptions of close contact, crowding, and careless hygiene matched those from other infected teams, and gave the coaching staff and epidemiologists a sense of how to break the chain of transmission. First, they swapped the liquid soap in the team showers for a strong anti-bacterial. Then, they required players coming off the field to shower before they went into the training room, barred them from playing with wounds unless the wounds were completely covered, prohibited sleeping in the locker room, disinfected the whirlpools, and raised the temperature of the water in the team laundry to a scalding 140 degrees. Nasal cultures showed that seven players, including one of the sick players from the previous year, were carrying MRSA in their noses. The team physicians required the colonized players to take oral and topical antibiotics to kill the bug.

For more than a month, the campaign worked. There were no more infections. Then, in late October, four more players came down with fresh boils, and one was hospitalized. The frustrated epidemiologists, who worked long hours during the week, gave up a Saturday afternoon to monitor a game. They spotted something they had never suspected. Despite all the training they had received about how to behave in the locker room, players and their student athletic trainers were sharing towels on the field. The coaches stopped the sharing, and bought the team a lifetime supply of tough paper towels. At last, the 2003 outbreak stopped.

In 2004, though, one previously infected player had a recurrence. After his diagnosis, the team toughened its rules even further. They cultured every player at the start of the season, required antibiotics if the cultures were positive, sprayed the weight room and treatment tables with disinfectant several times a day, and started carrying their own antibacterial soap to away games. There were no infections in 2005, or any year afterward. The Trojans appeared to have chased the bug for good.[16-18]

Over the next few years, MRSA infections tormented college and pro sports teams. In the NCAA, the victims included Mike Gansey of West Virginia University and Kenny George of the University of North Carolina-Asheville, who lost part of his right foot.[19,20] In basketball, Drew Gooden of the Orlando Magic.[21] In baseball, Sammy Sosa of the Baltimore Orioles and Alex Rios of the Toronto Blue Jays.[22,23] In football, Kellen Winslow and five teammates on the Cleveland Browns, and Brandon Noble of the Washington Redskins, whose pro career was ended by a persistent knee infection.[24,25] And at Lycoming College, twenty-one-year-old wide receiver Ricky Lannetti—who died from MRSA, hours before he was to start in the NCAA quarterfinals, and just four days after coming down with what felt like the flu.[26]

Pro teams with millions of dollars invested in their players spent lavishly to keep MRSA at bay. The Redskins, who had five other cases besides Noble, ripped out their locker room in the summer of 2006, repainting, replacing the carpet, swapping out shared benches for individual stools, and installing a $17,000 Jacuzzi with an ultraviolet light filter to kill germs in the water. They sprayed the new room, the training rooms, and the weight-lifting equipment with a costly coating intended to kill any germs that landed on it. And the large athletic training staffs hounded the players to keep up with basic hygiene: showering, using hand sanitizer, and keeping their towels and razors to themselves.[27]

But in schools, chronically squeezed for money, containing MRSA proved much more challenging. Which was why, on a hot morning in May 2008, Tom Keating, PhD, was standing in a high school bathroom in an Atlanta suburb with a look of distaste on his face.[28]

The bathroom's three stalls had no doors. The water in one of the toilets was yellow with urine, and the toilet-paper bracket in a different stall was empty and wrenched half-off the wall. Wadded-up paper towels covered the floor and had been crammed into one of the sink drains. The two soap dispensers were empty.

"And this is one of my better schools," Keating said.

It was the fifth sorry school bathroom he had seen that morning. Keating's first career had been as an education lobbyist in Atlanta, his second, briefly, as a school board member in nearby Decatur, Georgia. Now he was on his third—pursuing the quixotic goal of making kids into better citizens by giving them better bathrooms.

Keating was prompted into the third career by his own son and daughter. They routinely came home from middle school and high school complaining that their school bathrooms were vandalized and filthy; his son would beg the school nurse to let him use her private restroom instead. Keating started a one-man, largely self-funded advocacy initiative called Project CLEAN (for Citizens, Learners and Educators Against Neglect). Over time he worked with several hundred schools in eighteen states.

Keating estimated that at least a third of school bathrooms—about 330,000 across the country—were dirty, unhealthy, or unsafe. They got that way because, at some point, some kid would act out, breaking faucet handles or scrawling on the walls. But they stayed that way because there were no federal and almost no state regulations that set any standards for how school bathrooms should be kept. Municipalities were cutting school budgets so deeply that custodians were overwhelmed, and teachers were forced to focus so tightly on federally mandated test results that they had no time and little enthusiasm for monitoring restrooms. And when the restrooms weren't cleaned up, they got worse. A clogged sink would soon be followed by unraveled toilet paper, kicked-in stalls, and puddles of urine on the floor.

In a few school districts, Keating parlayed his research into ongoing contracts to monitor bathrooms, talk to students, and persuade administrators to take up the cause. "In an average school in America, four out of ten sixth to twelfth graders avoid the bathroom," he said, approaching a group of girls slouched against lockers near the bathroom door. He asked them if they ever used it. They gave him a look that vaulted past practiced teenage pity and landed somewhere between outrage and disgust.

"I *never* use the bathroom here," one said. "My mom doesn't understand why I run straight to the bathroom when I get home. She's like, 'Come tell me about your day!' and I'm running up the stairs."

"I use it when I have to," her friend said. "But I never use the soap. Because the soap in there, it's clear, and you don't know if it's really soap or spit or what. We used to have pink soap; I wish we could have that back."

Keating had started his project to honor children's needs and to improve their schools, but he had heard enough stories from students about soap, and enough public-health messages about hand washing, to wonder whether his advocacy had relevance for MRSA as well.

"Soap is an issue in maybe 40 percent of the schools I go to," he said. "Soap dispensers are broken, or they're not filled, or they're not even there because of vandalism. Kids go to the bathroom, or they go play sports, and they can't wash their hands."

In Stafford, Texas, high school trainer Dave Edell was convinced of the connection. In the five years since Nick Johnson fell so gravely ill, there had been only two more serious cases of MRSA at Stafford, neither of them as severe as Nick's, though every so often, kids would get minor boils and rashes that would prove to be MRSA. Success in quelling the bug may have come from keeping such a close eye on it, a luxury that larger school districts could not afford. Stafford's schools are so small and tightly knit that every coach and teacher knows every student's extracurricular activities—so if a girl athlete developed a skin infection, Edell would know that she also played in the marching band, and could tell the music director to send her uniform for dry-cleaning until her doctor cleared her. The school's housekeeping staff and the athletics department had tight ties too—the director of maintenance is married to the girls' volleyball coach—so it was simple to pass on the message that the locker rooms were overdue for major cleaning. The custodians and the trainers and coaches all kept their noses tuned for smelly lockers, and periodically would compel the kids to haul out filthy workout gear and take it home to be washed in hot water, while the staff sprayed shoulder pads and any shared equipment with diluted bleach.

Stafford had never performed what Edell irreverently called a "nuclear cleaning," the disinfectant bombing and spray-coating that had become a huge business since pro teams began disclosing their outbreaks. The CDC advised against it, arguing that any expensively cleaned locker room could be recontaminated the next time a colonized kid walked in.[29] Instead, Stafford worked to educate its students. All the coaches talked to their teams about MRSA, and the school hallways were papered with earnest posters: "A Good Offense Is Still the Best Defense! Here's Our Plan to Sack Staph."

The cleanliness message seemed to be getting through. On a weekday morning late in 2008, while Edell was icing one girl's knee and applying an ultrasound probe to another girl's strained calf muscle, one of the high school softball players bounced into the training room.

"May you clean my blister please?" she asked.

Edell sighed, peeled open a bandage and a single-use tube of ointment, and looked at her over his metal-framed glasses.

"You're supposed to wear a batting glove," he reminded her.

"I left it at home," she said, tossing a ponytail of narrow dreads over her shoulder. "And my blister bust. But clean it good, 'cause I don't want no staph."

Edell was pleased to hear what sounded like awareness. The school had worked hard to achieve it, and the athletic staff made sure that soap, hand sanitizers, and bandages, and adults to help apply them, were always available to reinforce the message.

But no matter how scared of MRSA the kids were, and how much Edell, the coaches, and the school worked on them, there remained one fundamental thing they simply would not do: they would not shower at school. Edell found it astounding. He came from an athletic family;

both brothers and both sisters played varsity sports. He had been either in school himself for advanced degrees, or working in schools, for practically his whole life. Showering after sports was something he took for granted.

"There is not a single one of those girls who will come in from practice, go into the locker room, shower and go to class," he said. "They wash their hands and face, maybe they wipe off with a towel, and that's it. My high-school football players, they change, go out to eat, and maybe they shower when they get home. They think they're not dirty, but they don't understand that they've had the person-to-person contact that causes staph to spread."

The Stafford kids were not unusual. Over the 1980s and 1990s, without anyone making a ruling or taking an official position, school showers quietly slipped out of teen culture. A 1994 court case in Pennsylvania may have contributed. An overweight girl who feared she would be tormented by classmates sued to break a mandatory-showers rule.[30] (The suit, brought by the American Civil Liberties Union, was withdrawn after the school district made showers optional.) But in some states, showers stopped long before that. States cut budgets, so schools could no longer afford maintenance or towel laundries. Students loathed gym classes, so schools jazzed them up to lure kids into exercise—building rock-climbing walls, buying inflatable balance balls—and de-emphasized the humiliating, privacy-invading showers.[31-34]

In one of the Georgia schools where Tom Keating worked, the girls' pink-tiled showers were bone dry and the faucets were tied off with plastic tape. "We don't shower," said a student who was walking through to the toilets after basketball practice. "Not ever. We haven't used these in years."

Dave Edell and Nick Johnson's mother, Janet, had never been able to determine what caused Nick's MRSA infection. With no evidence but MRSA's known behavior, their best guess was that the tackle had driven his shoulder pads into his skin, creating an invisible abrasion that opened the door to infection. Privately, Janet wondered whether schools' letting kids slide on cleanliness had contributed. She had been shocked to realize that mandatory after-gym showers, a staple trauma of her own Baby Boomer adolescence, had never been a part of her child's education. She envisioned a resistant staph strain lurking on his skin. Maybe it had been there a long time. Maybe he had picked it up on the day he made the tackle, crashing against his classmates in the muggy Texas afternoon.

Maybe, if he had showered, if she had insisted he shower, it would have made a difference.

"The kids say they don't have time, but we did it, and we had the same twenty-four hours a day they did," she said, paging through a scrapbook that documented Nick's illness and recovery. "And the schools don't make them do it. I think they should do it again. I think it would protect the kids that much more."

School children opted out of showers out of embarrassment, and ignored basic rules of hygiene out of laziness or lack of knowledge. Like teenagers everywhere, their belief in their own invulnerability kept them from realizing the risks they exposed themselves to. But in another type of institution, a different segment of the population was also going without basic hygiene, and this time by force, not by choice. In jails and prisons, institutional ignorance, underfunded medical care, and overcrowding all restricted the preventative measures that kept MRSA at bay in sports teams. The bug was burgeoning there, among prisoners and among those who cared for them—and thanks to the constantly churning jail popula-tion, and to unknowing prison guards and workers, MRSA was moving back out into families and neighborhoods as well.

CHAPTER 9

PRISON, THE PERFECT
INCUBATOR

Kevin Keller lay on his back on the bunk that jutted out from the wall. He looked at the cell around him. There wasn't much to look at: solid door, bare walls, a metal toilet. It was Labor Day weekend 2002, and it seemed quiet outside his cell. It was hard to be sure, though, because his head was ringing with the fever that had gotten him put into this solitary confinement late on Saturday evening.

Keller had been trying to get medical care for weeks, starting one month after he landed in the Bucks County Correctional Facility in Doylestown, Pennsylvania. He knew the jail stay was his own fault. Back in July, he had taken his father's car without permission, which would have been a minor infraction, except that Keller's license was suspended because he was on parole for driving drunk. The police pulled him over. He went to jail for the mandatory 180 days that a parole violation earns.

By jail standards the Doylestown jail was not a bad one. It was seventeen years old, and had replaced a Victorian-era monstrosity that had been turned into an art gallery. Though he was only twenty-five, Keller knew the new jail well. He'd been inside before, jailed for drinking and drug offenses since he was a teenager. This time, though, he had been paired with a cellmate who had an oozing boil on his back. Soon Keller had picked up an infection as well.

Wincing at the effort, Keller levered his torso off the cot, pulled up his knees and pivoted so he could set his feet slowly on the floor. He carefully stood up, got his balance, and shuffled over to the toilet. He unzipped his prison-issue jumpsuit, letting it fall awkwardly to the

125

ground, and pushed the prison boxer shorts down to join it. He settled on the toilet, exhaled and looked down. His scrotum was puffed and hot, as big as a grapefruit and as dark red as raw meat, and he could see the open sore that had been weeping pus for six days now. Sliding his hands around his testicles, Keller clamped his eyes shut, gritted his teeth, and squeezed.

When the white-hot agony faded and he could breathe again, he staggered erect, pulled up the shorts and the jumpsuit, and sagged back on the cot. He thought he might have two hours before his scrotum swelled so painfully that he would have to do that again.

The prison had said he could see a doctor on Tuesday. That was two days away. He wondered if he could make it that long.[1]

From the moment the new Bucks County jail opened in 1985, MRSA had been a problem. Though the facility was new, it was built without a dedicated infirmary. To service the ninety-six inmates on the women's side and the 641 inmates on the men's side of the jail, there was a dispensary that was open only during the day. When inmates complained about something that looked like a contagious disease, they went into solitary confinement because there was nowhere else to isolate them. So no one on either side of the jail admitted to illness until they needed care badly, because no one wanted to go into what inmates called "the hole."

Starting in the middle of 2000, two Doylestown attorneys named Anita Alberts and Martha Sperling began working on a court-ordered investigation of conditions for women in the jail, which only a few years after its opening was holding many more prisoners than it was designed for. Their assignment was determining whether the jail had adequate mental health facilities for its relatively few female inmates. Soon, though, they started hearing stories about physical illnesses as well. At least fifty inmates—approximately half of the women being held there—told them almost-identical stories about MRSA infections and about the conditions they believed were causing them: shared bathrooms and showers that were sloppily cleaned by inmate trustees; ripped, stained mattresses infested by insects; and laundry that never seemed to get clean. "They would show us these rashes, abscesses the size of golf balls; they couldn't put their arms down," Sperling recalled. In December 2001, as their report was about to be published, forty-two-year-old inmate Virginia Brejak died of a stroke that her family believed was caused by a MRSA abscess. The family and her fellow prisoners attested that for months beforehand, she had been complaining of boils on her head and body and saying that she was not getting the medical attention she needed

The inmates' complaints were so serious and similar that the Pennsylvania Department of Corrections and the state Department of Health sent representatives to inspect the jail. In separate reports, the agencies said that the surroundings were dirty, that inmates could not be segregated when medically necessary, and that there were twenty current cases of MRSA infection. In September 2002, Alberts and Sperling filed suit on behalf of four female inmates. They knew it was a long shot—few people would consider petty criminals to be innocent victims—but they charged that the women's rights were being violated by the threat of disease and the reality of inappropriate or neglectful treatment.

The lawyers didn't know that on the men's side of the building, Kevin Keller was enduring the same neglect.

About a month after he was admitted, Keller was showering, and felt something under his arm. It was a red pimple, conical and hard, rising in his armpit. After a day or two, it popped, but then more pimples came—always at least one, but usually four or five at a time. He dismissed them, thinking that maybe the jail soap was bad, or the deodorant they let him have was irritating his skin.

A portion of the jail's inmates, ones who had been given short sentences or committed less serious crimes, were considered low security risks, and given jobs to do around the jail. Keller's job was in the kitchen. Women prisoners worked there too, and Keller kept noticing how many of them had bandages on their arms or faces. The jail's gossip network buzzed with news of MRSA. Keller had no idea whether that was what he had, but he knew he'd never seen anything like his boils before. Several times he wrote his name onto the list for sick call, and waited to be escorted to the infirmary. He was never called. Once, coming back to his cell at the end of the day, he spotted that morning's list on the floor of the common area with a footprint on it.

After about two weeks, the infection moved down his body. A small eruption on one testicle grew larger and larger, until it came to a head and broke open. It was tender and wet and smelled. "Like decaying flesh," Keller said. "I will never forget that smell."

Even with the weeping wounds, he could not convince the jail staff to let him off work for sick call. On the Saturday before Labor Day, his girlfriend came to visit, and they talked in the day room where detainees were allowed to meet. When she left, he spotted his chance and slipped away around the corner. He sneaked to the dispensary, where a nurse's aide was closing up for the weekend. "I've got a situation, and I need help," he said. He dropped his trousers.

The guards threw him in solitary.

They agreed that he needed to see a doctor, but there was no arrangement for calling in a physician over the weekend. The next scheduled doctor visit would be Tuesday. It was the last conversation he had with anyone for forty-eight hours. Over Saturday night, Sunday, and Monday, he lay on the metal bunk, feverish and frightened, feeling his genitals swelling with infection until he was afraid they would burst. On Monday night, a nurse who had just come on shift to assess the medical needs for the next day checked on him and ordered him to be transported to the local hospital immediately. He was there for ten days, on IV antibiotics. When he was released, the jail put him back into solitary again.

There was a surgical incision down the center of his scrotum, from where the hospital had slashed the infection to let it drain. Keller was supposed to soak it in hot water and keep the wound covered and clean. But there was no hot water in his cell, and the jail confiscated the protective cup the hospital had given him. He asked for bandages, but they gave him only little Band-aids, so he packed the wound with toilet paper instead.

Fearing he was an infection risk, the jail volunteered to discharge him, and he was released a month early in November 2002. His family pressed him to file a lawsuit, and since Doylestown is a small town, Keller soon found his way to Alberts and Sperling. The lawyers were still collecting accounts from prisoners, working on expanding the women inmates' suit into a class action lawsuit covering both sides of the jail. They told Keller about another male inmate who had fared just as badly. He'd been in the jail the year before Keller, had a leg abscess, was denied care and developed MRSA osteomyelitis, the severe bone infection. The doctor who treated him described his leg after lifesaving surgery as "filleted."

In July 2003, Sperling and Alberts filed suit on behalf of Keller and the other inmate, Benjamin Martin. In January 2005, a federal jury awarded them $1.2 million.

Almost as soon as the community strain of MRSA emerged, it became clear that jails and prisons would play a special role in sustaining the strain and spreading it.[2] In November 1999, a few months after the CDC published Tim Naimi's account of the four children in Minnesota and

North Dakota, officials at a state prison in Mississippi began noticing an unusual number of skin infections among the 1,800 men and 1,200 women incarcerated there. They always expected a few infections, from poor health or poor hygiene, but the numbers now seemed high. By the following October, fifty-nine inmates had fallen ill: twenty-seven with noticeable boils or abscesses, twenty-one with cellulitis, and eleven with open, weeping wounds. Two were sick enough to be transported out of the prison and hospitalized.[3]

Around the same time, a minimum-security detention center in Georgia was experiencing similar outbreaks: a cluster of eleven serious skin infections among the two hundred men who lived there, then another cluster of fourteen cases, and then five more.[4] At the other end of the state, seventy inmates in a maximum-security prison also developed serious infections. So did prisoners in institutions in a half-dozen other states—but there was no reason for those jails and prisons to talk to each other, and no single surveillance system that would have rung an alarm bell over so many small outbreaks at the same time.

And then, in the fall of 2001, inmates in the Los Angeles County jail started complaining of spider bites.

It was not unreasonable that there be spiders in the jail. Southern California is warm enough to sustain almost any insect, and the jail, a hulking complex that squats south of Dodger Stadium, is large enough to hide entire civilizations of bugs. By population, L.A. County is the largest jail in the country, with at least 20,000 inmates at any one time. At least 600 inmates arrive every day, and more than 165,000 move through in the course of a year. The Los Angeles Sheriff's Department, which operates the jail, brought in exterminators to spray pesticides and search for hidden infestations. The inmates' bites, small, hot pimples, were blowing up into significant skin infections, and some of the victims had to be hospitalized. The jail's health services began giving out anti-inflammatory drugs and antibiotics, and culturing the lesions to have a record of whatever was going wrong. Still, the skin problems continued, and in frustration the exterminators caught a few of the spiders and sent them to the county entomologist—who called them back, perplexed, to tell them that their spiders belonged to a species that do not bite.

That discovery ruled out the explanations they had been hoping to hear: either that the spider was something like a brown recluse, which gives a bite that swells up like a pimple and breaks down into an ulcer, or that the spiders were accidental carriers of some bacterium that their bite inserted into the skin. Mystified, they turned back to the culture data they had been gathering for almost a year now, and found something they had not noticed: the jail was the site of the largest MRSA outbreak yet seen in the United States. Just in 2002, 928 inmates had had infections, 66 of them serious enough to require hospitalization.[5] As the outbreak continued, the total would climb to thousands more.

The sheriff's department called in the Los Angeles Department of Public Health, which handed the investigation to Dr. Elizabeth Bancroft, a dynamic epidemiologist who had trained there as a CDC disease detective and stayed when her term was up. She quickly saw that everyday life in the jail was an extreme expression of the MRSA-fostering conditions that CDC investigators had found in sports teams, a set of factors they were calling the "Four Cs": crowded conditions, contaminated surfaces, insufficient cleanliness, and skin-to-skin contact.

But jails, she realized, had particular complications. They were overcrowded; L.A. County always seemed to be double-bunked, which meant there were no spare cells to use for isolation. There was not enough clothing to go around—generally, prisoners got new underpants twice a week and a new jumpsuit once a week—and limited access to the showers. There was no guarantee that laundry would be washed at temperatures hot enough to kill pathogens, or that surfaces would be cleaned with bacteria-killing soap. Alcohol hand gel was ruled out because it could be flung in guards' eyes or burned. In-house medical clinics were overloaded and there were long waits for appointments. Plus, the clinics charged a co-payment; so inmates who had no credit in the jail accounts resorted to self-doctoring, lancing boils with whatever sharp objects they could find and spreading infection because they could not adequately clean or bandage the wounds they made.[6]

"I don't know if we can get rid of it from places like this that have these kinds of crowded living conditions and suboptimal hygiene," she admitted. "Just as we've never been able to completely get rid of scabies, or lice. At best we can probably identify it, treat it, try to control—but I don't think we'll eliminate it, unless it dies out on its own."

People who are admitted to jails and prisons are much more likely than the general population to have been homeless, malnourished, infected with sexually transmitted diseases, drug-addicted, or suffering from chronic illnesses that undermine their immune systems. Overcrowding forces closer contact between the infected and the immunologically vulnerable, and the constant churning of the population practically guarantees that new, uninfected inmates will be exposed. Distrust between jail inmates and guards means that any request for an infirmary visit may be interpreted as the first step in an escape attempt. And safety regulations make it impossible to reserve equipment such as stethoscopes and blood-pressure cuffs for infected inmates, because if they are left in cells unsupervised, they might be converted into weapons.[7]

All those factors made jails and prisons practically perfect incubators in which MRSA could breed, spread, and evolve. Rates of infection and colonization in correctional institutions began to climb far beyond what they were among the general population. In San Francisco in 2002, before USA300 had firmly entrenched itself, MRSA already accounted for 75 percent of all staph in the county jail, surpassing the rates in hospitals, even in intensive-care units.[8] Outside prisons, the proportion of the U.S. population that was colonized with MRSA was 1.5 percent, but it was 4.5 percent in a Texas county jail, 10 percent in New York State's maximum-security prisons, 16 percent in the Baltimore city jail, and 17 percent in New Orleans.[9–12] Meanwhile, the U.S. incarcerated population was growing, reaching 2.3 million in June 2008. Every month, more men and women entered jails and prisons, and more and more of them acquired MRSA while they were there.[13,14]

MRSA's advance left correctional institutions grappling with more disease than they had training or staff to deal with. They jury-rigged solutions that curbed the little liberty and autonomy inmates have, without appreciably improving infection control, as demonstrated in the case of Ohio inmate James Bell.[15]

Bell, who is in his early sixties, has been incarcerated in Ohio since 1984. He was originally sentenced to life in prison for the murder of a man who he says robbed his wife and stepchildren. In April 1993, he was a prisoner at the Southern Ohio Correctional Facility in Lucasville, Ohio, when an eleven-day riot broke out. Initially Bell served as a prisoners' spokesman in negotiations over guards taken hostage, but he was later accused of beating a guard. A year later, he was convicted of assault and had twelve years added to his sentence.[16]

Bell, who converted to Islam in prison and took the name Abdul-Muhaymin Nuruddin, developed kidney disease in 2004, when he was fifty-seven. It quickly became severe enough for him to need regular dialysis, and the prison medical system arranged for him to have a dialysis graft, an artificial connection between an artery and a vein that allows easy access for the catheters that carry blood away from the body and back, after it is cleansed of toxins. By 2007, he was receiving dialysis three times a week.

For years, medical care in the Ohio prison system had been contentious. In 2003, three inmates—two with hepatitis C and one with severe gum disease—sued the state, alleging that they were persistently denied access to health care. Their suit became a class-action suit on behalf of the 44,000 prisoners in the state, and a court-appointed medical investigative team began visiting prisons around the state to assess the conditions. The team was so disturbed by its findings that they cut short their tour after visiting eight of the planned twelve institutions, and filed a scathing report. It led to a January 2005 settlement agreement that reorganized spending, staffing, and planning for infectious diseases in Ohio's prisons.[17]

Left unmentioned in the settlement—though discussed in the medical team's report—was a 2003 outbreak of MRSA in the Pickaway Correctional Institution in Orient, Ohio, east of Columbus. Eighty inmates there had complained of spider bite–like lesions, and were diagnosed with staph infections and given penicillin. The infections, though, were MRSA—but this was only revealed after nineteen-year-old Sean Schwamberger, serving eleven months for forgery, collapsed and died of MRSA toxic shock. Another inmate, Clarence Grubb, died of a MRSA infection just three days after being discharged. After their deaths, the Ohio Department of Rehabilitation and Correction began requesting and recording MRSA cultures of prisoners' wounds. In eleven months, they found almost five hundred.[18-20]

The 2005 settlement called for additional physicians, nurses, patient management, and infection control in the Ohio prisons, but it did little to stop MRSA from occurring or spreading. In the summer of 2007, other inmates in the block where James Bell was housed began complaining of spider bites, the same small, hot pimples that had heralded MRSA in prisons around the country. Bell never had a bite, but in October of 2007, the dialysis graft in his right thigh became hot and swollen, and he began to feel feverish. One day, he passed out during dialysis. He woke up at Ohio State University Hospital in Columbus, twenty miles away, under protective custody, with a 103-degree fever and an IV in his arm. MRSA had invaded the graft and, because it provided direct access to his circulation, had immediately caused bacteremia. He was so sick that the dialysis graft in his leg had to be removed, and another graft created on his other leg.

Because of his need for dialysis, Bell had earlier been transferred from a high-security prison to the Correctional Reception Center, a mostly lower security institution not far from Columbus and in the same town as the Pickaway prison where Schwamberger died. After thirty-three days in the hospital, on IV antibiotics and mupirocin for decolonization, he was brought back to the Center. But he was now known to be a carrier, and therefore considered an infection risk for the rest of the jail—plus he was a maximum-security prisoner in an institution designed to hold few prisoners in that category. He was put in solitary, on lockdown twenty-three hours a day.

The restrictions may have protected other inmates, but they did little for Bell, who struggled not to contaminate his surroundings. His cell could be cleaned only when he was not in it, and if the trustees were busy when he was escorted to his hour of exercise, it would not be cleaned at all. Antibacterial soap, lotions, and first aid supplies were available only if he bought them with jail credits through the commissary. As in other prisons, alcohol hand gel was considered contraband.

In November 2008, Bell's MRSA recurred, and then became persistent. Repeatedly he was treated at the hospital and the nearby Corrections Medical Center. His dialysis graft kept getting infected, and doctors moved it three times—to the other leg and then to each arm. When he ran out of usable veins, doctors began using an artificial dialysis connection called a port, a plastic tube that they stitched into blood vessels in his chest and groin. Thanks to recurrent MRSA, each site failed in turn, making lifesaving dialysis increasingly difficult to perform. Between the original infection and 2009, Bell underwent eight surgeries to create grafts or move dialysis ports.

"I may not be able to receive another dialysis graft," he lamented in early 2009. "Then life for me sadly comes to an end."

Most prisoners are not like James Bell. They serve their sentences and leave—or, increasingly, are released early due to overcrowding. More than 2 million Americans are in jail at any moment, but more than 13 million cycle through jails and prisons in a year. In 2006, approximately 1.5 million people left correctional institutions carrying an infectious disease.[21]

For a very long time, no one wondered what the effect might be of so many prisoners picking up MRSA in jail and carrying it with them when they are released. In 2006, though, a group of epidemiologists at the University of Maryland became interested in a vast increase in MRSA in the main jail in Baltimore and wondered what its impact on the surrounding city might be. They used a measure called "epidemiologic weight," a calculation that assesses how much risk an outbreak in a self-contained group poses to its larger setting, based on how many people join the group, become infected without receiving treatment, and exit the group, as well as how fast or slowly that churning happens. Because hospitals have hundreds of patients staying for weeks, while jails have thousands of inmates staying for years, the researchers assumed that hospitals would feed the larger epidemic, while prison outbreaks would harm only those inside. Their calculations revealed that assumption was wrong. Jails and prisons had a greater epidemiologic weight than hospitals—though hospitals had been in the spotlight for decades as MRSA-infection risks, while jails had received almost no attention at all. Jails and prisons, they said, were "superspreaders," seeding MRSA into society at large.[22]

The concept helped explain something that MRSA researchers had been worrying about. In several cities, the bacterium was not evenly distributed: infection and colonization clustered in certain neighborhoods. That was especially noticeable in the more than one hundred public health clinics in the Chicago metro area operated by John H. Stroger Jr. Hospital, the successor to the city's iconic Cook County Hospital, the model for the TV show *ER*. Using technology that was not available when Robert Daum was recording the first USA400 cases a decade earlier, researchers at Stroger took the medical records of patients with community MRSA infections, linked their diagnoses to their addresses, and plotted them on a computerized city map. The densest clusters matched neighborhoods where, according to census data, residents were more likely to be poor, to be living in public housing, and to have a family member who was incarcerated. (A separate study in one neighborhood found that 29 percent of residents either had been in jail themselves or had a family member currently incarcerated and expected to return.) The residents were also more likely to be transient, suggesting that the infection that someone acquired in prison and brought home to his family on release might travel with the family to a new neighborhood when they moved.[23]

"I personally believe that jails and prisons are primary community incubators of MRSA," said Gregory Belzley, an attorney in private practice who became so outraged by conditions in small-town jails that he filed seven class-action cases on behalf of Kentucky and Indiana inmates.

"The football teams, they've got the money to deal with this problem. The hospitals, they've got teams of infection control experts breaking their butts to solve it. But I walk into a jail and there's sixteen guys sharing a cell built for eight, one sink, one toilet, two of them have draining sores, and no one is taking responsibility."[24]

MRSA in correctional facilities affects a wider segment of the population than prisoners, and reaches neighborhoods far beyond the ones where former prisoners live. About 750,000 people go in and out of correctional institutions in the United States every day. They are administrators, guards, cleaners, and cooks, and they have enough close contact with prisoners to be vulnerable to their infections, as well.

Chris Pearson, a forty-three-year-old correctional officer at Folsom State Prison, thought it was odd when inmates began to complain about spider bites. In the twenty-two years he'd worked there, he'd never noticed spiders being much of a problem.[25]

Pearson had been an officer at "Old Folsom," the sprawling property made famous by Johnny's Cash's 1960s concerts, since getting out of the Marines when he was twenty-one. He worked in Ad Seg, the secure "administrative segregation" unit where every one of the prisoners had to be strip-searched and handcuffed before being moved out of a cell. Even with latex gloves on, skin contact with inmates—or their laundry or their clothing, soiled and sweaty from daily breaks in the prison yard—was inevitable.

The guards suspected there was MRSA in the prison. They heard prisoners talking about bites and boils, they escorted them to and from sick call, and they kept an eye on the sores that so many inmates seemed to have. That was challenging to monitor, because Folsom was so crowded. The entire prison was at twice its official 2,000-bed capacity, and Ad Seg, built for 92 men, routinely had more than 150 crammed in. Plus, medical care in all of California's prisons was suspect. The state had successfully been sued in 1995 over inadequate prison mental-health care, the same wedge issue that had gotten Anita Alberts and Martha Sperling into the Doylestown jail. And the entire prison medical system had been in court-appointed receivership since 2005 thanks to the judgment in a second lawsuit that found health care gravely inadequate.[26]

Pearson's union, Unit 6 of the California Correctional Peace Officers' Association, had been arguing with the prison's administration since at least 2004 over their suspicions that MRSA was spreading through the cells. Officers wanted to be informed when prisoners were diagnosed, so they could sanitize equipment, make sure cells were scoured, and take extra care that they were not exposed themselves. The prison leadership had refused, saying that it would violate the inmates' right to medical privacy. The guards believed there was more MRSA at Folsom than anyone was admitting. The "spider bites" the inmates complained about were treated at the infirmary with the same antibiotics used for MRSA, and the guards wondered whether the administration hid behind its privacy argument to keep from admitting how big the outbreak might be.

On a Tuesday morning in April 2006, Pearson woke up with his nose hurting from the inside, a feeling so odd that he thought it must be a cold starting, or a pimple in a place where he'd never had one before. By Thursday it was inflamed—his wife told him it looked angry—and so sensitive that merely flapping his hand in the air in front of his face made it hurt. By Friday, he couldn't breathe through one nostril. He saw a nurse practitioner over the weekend, but by Tuesday morning his whole face had puffed up and he couldn't open his left eye fully. He drove himself to the emergency room, where they listened to his symptoms, did a quick test, and immediately stashed him in an isolation room with a clindamycin IV. He had MRSA in his nose and sinuses, unnervingly close to his brain. They gave him a steroid to bring down the swelling, and after a few hours, he could breathe freely and open both eyes again. After twenty-four hours, they sent him home with a ten-day course of antibiotics and orders to stay out of work for a week.

Pearson's infection was briefly frightening, but his colleague Gary Benson's was life-altering. Benson was fifty-five, a former communications engineer who had been working in the California prison system for seven years. He had just gotten divorced, and was working extra shifts to build up his finances. His main job was monitoring inmates committed to Folsom's minimum-security drug rehab program, but his additional work took him into the regular cell blocks and the Ad Seg unit.

In July 2006, he started to feel achy and run down; he would have called the symptoms flu-like, but flu season was long over. He was nauseous and sweaty. His joints ached and his groin was inexplicably sore. He went to his doctor and got checked for flu and for anything that might make his lymph nodes swell, but nothing showed on a flu test, and there were no lumps or red patches to indicate anything wrong.

On his third morning feeling ill, he woke up confused and feverish. He pushed down the covers and sucked in his breath: his scrotum was the size of a cantaloupe. He rushed to his doctor's office, and they got him to the ER. The resident who examined him took one look, told him it was MRSA, paged the surgeons, and ordered a bag of vancomycin. The doctors told him he needed surgery, but first they wanted to make sure he had an adequate dose of antibiotics in his system. A few hours later, two nurses came in, gowned and gloved in accordance with isolation orders. The lab had run an antibiotic sensitivity test, they told him. They had tried eleven drugs against his MRSA strain, and nothing had worked. They were running the test again, to make sure that vancomycin would work.

"I joked, a bit," Benson said. "I said, 'What do you mean, I'm going to die?' But she wasn't joking. She said, 'If this doesn't work, there may be nothing we can do for you.'"

The test reported that the strain was vancomycin-sensitive. Early the next morning, as Benson was being prepped, the surgeon who would be working on him came by to explain what would happen.

"He said, 'We'll drain your scrotum, and then we'll see how much damage there is,'" Benson recalled. "If there's damage to a testicle, we're going to remove it. If there is damage to both, we're going to remove both, and probably your scrotum too.'"

When he woke up after surgery, everything was still there.

He remained in the hospital for ten days, and was bed-ridden at home for a month on IV antibiotics. Two times a day a nurse came to administer drugs and repack his wound. Benson was forced to miss work for a second month. Persistent pain nagged him, and in 2007 a urologist recommended a second operation to remove a duct that the infection had damaged. He was out of work for six weeks for that surgery, and the pain never diminished. At the end of 2007, he was declared partially disabled, though not enough to fund a full retirement. So he continued to work at Folsom in the prison's emergency medical clinic, where he saw at least one inmate come in with staph every day.

"I have to work, so I have to be around it," he said. "It worries me. They said I am ten times more likely to get it again, and if I get it, and it becomes vancomycin-resistant, then there might not be anything that would stop the infection. And that would kill me."

Benson did not know it at the time, but when he first came to the ER his infection made a deep impression on one of the doctors treating him. Two weeks after performing Benson's surgery, urologist Dr. Kaushik Desai wrote a letter to Folsom's warden. "I feel that I should bring to your attention that there might be a situation in your system that . . . is increasing the risk for your officers," the doctor wrote. "It may need the attention of an infectious disease specialist to work with the prison system . . . so you can prevent spread of this resistant bacterium."

Desai's concern was well placed. The bacterium was spreading, not just within the prison, but beyond its walls. David White, a Folsom veteran like Chris Pearson, worked in what the prison calls "the camp," a low-security unit that is the last stop before inmates finish their sentences and return to the real world. Prisoners there live in fifty-man dormitories, do cleaning and landscaping, and work in prison factories. Looking toward retirement, White had been picking up overtime shifts around the rest of the prison, and in late summer 2006 he was offered some extra hours in the drug-treatment facility where Benson had worked as well.

White had been working in the yard around his house, ripping out underbrush, and had gotten a case of poison oak. In the drug wing, someone spotted the rash on his arm.

"You better be careful," White remembered the guard saying. "They've got like ten cases of MRSA here."

White hustled for bandages, swearing—he knew the risks of an open wound, but he had not known there was MRSA on the unit. A few days later, on vacation with his wife, he spotted welts rising on the same arm, an inch high and several inches across. He resisted cutting short their trip, but she insisted, and they drove home and went to their doctor. The doctor confirmed that the welts were caused by MRSA, from a very resistant strain. The physician opened the abscesses surgically, wrapped White's arm, and gave him oral antibiotics that the bug had tested sensitive to. White went home under orders to isolate himself at home until the sores healed.

White called in to work, and was told to take six weeks off. It seemed absurd to him—the wounds were wrapped and he didn't feel sick—but it turned out to be fortunate, because several weeks later a boil began rising in his wife's groin. It was clearly MRSA. When they got her to the hospital, tests proved it was the same strain as his, the strain that he believed he contracted in the drug unit at Folsom. The symptoms it caused in his wife were much more severe. She was hospitalized for four days with bacteremia, and was discharged with a catheter stitched into her chest to accommodate home treatment with a vancomycin IV.

For a week, a nurse visited two times a day to change IVs and wound dressings. And for two months all told, the Whites went nowhere but the doctor's office. Their physicians ordered them isolated at home. They were not allowed to shop, to socialize, or to have anyone over. They cancelled their Thanksgiving dinner. They were specifically instructed to avoid vigorous exercise until tests proved that the bacteria circulating in their bloodstreams had cleared. Dave White, accustomed to working outside and keeping busy with projects, felt he had been thrown into his own form of solitary confinement. But the cautionary measures were prescient: his wife's infection recurred. They were not declared clear until late November.

The Folsom MRSA strain was virulent and multidrug-resistant, and caused an epidemic that persisted for years. But it was presumed to be a low risk for the outside world because what happened within the thick stone walls of Folsom was thought to stay there, just as the inmates did. The illness of White's wife showed how wrong that assumption was. The strain would not stay confined to Folsom; it could walk out of the prison in the nose or on the skin of anyone who had visited.

The correctional officers had given up attempting to persuade the prison of the risk. They filed a complaint over MRSA cases with Cal OSHA, California's work-safety agency. In October 2007, the agency found in their favor, fining the prison almost $21,000 for not reporting or trying to reduce MRSA cases among the guards and ordering it to develop a MRSA prevention program. The Department of Corrections appealed, freezing the Cal OSHA finding and the corrective actions it recommended until the appeal process was exhausted. As the legal battle raged on, another guard, Alma Zavala, died in 2008 of MRSA pneumonia and a cerebral hemorrhage.[27] Anxious to force action, the guards' union sued the state. In mid-2009, it still had not gone to trial.

MRSA had burgeoned in jails and prisons without anyone but a few doctors and correctional officials, and the prisoners themselves, realizing it was there. It was about to do the same in another confined population, one even less able to speak on its own behalf: farm animals. Once again, the bug displayed its extraordinary talent for adaptation, subtly altering its genetic makeup to fit into an ecological niche where it could reproduce undetected. And once again, it would demonstrate its unmatched ability to cause new forms of illness, reaching out through livestock to threaten farmers, and then farm neighbors, and possibly the food supply.

CHAPTER 10

INTO THE FOOD CHAIN

It was cool and dim inside the old swine barn. Humidity lingered thick as mist in the corners of the thick brick walls. A thunderstorm had roared through the night before, the first harbinger of autumn come unfairly early, and two hours after dawn, clouds still hung low over the fairgrounds. Outside the barn, early arrivals milled in lines for cinnamon mini-donuts and slabs of maple-glazed bacon on a stick, fueling up for the gleeful, raucous chaos of Labor Day weekend at the 2008 Minnesota State Fair. Solemn high schoolers in rubber boots worked over their pet llamas with hair dryers and brushes, keeping a nervous eye on the ranks of exhibition judges in fleece vests and tractor caps. In an asphalt lane lined with trailers and humming generators, two women chatted and smoked while the horses they were leading nibbled their shoulders and snorted into their hair.

The swine barn smelled of burned coffee and bologna-and-mayonnaise sandwiches, and ammonia rising from the wet sawdust where the pigs had been lying. The fair had been going for a week, and most of the pigs were gone, sold for slaughter or trucked home for breeding. Most of the slatted-wood enclosures stood empty, breed banners and tinsel streamers hanging limp over upended feed pans, wire gates left unlatched for the cleaning crews that would scour the concrete floors before the last class of pigs and pig-raisers moved in for the weekend. Near the door, 1,240-pound Squeaky, "Minnesota's largest boar," snored next to a table of pig-shaped key chains.

In a corner, two eight-month-old sows destined for auction slept under placards that advertised their grand-champion awards and specified their owners and breed. Dr. MacDonald Farnham, a veterinarian from the University of Minnesota, snapped on a pair of bright-blue gloves and unwrapped a set of swabs that resembled long, single-ended Q-tips. Squatting next to the pen, he reached a long arm under the gate, delicately inserted a swab into one pig's nostril, and quickly twirled it. The pig snorted and shook itself awake, stumbling to its feet and backing away. Farnham quickly drew the swab back, deftly avoiding the pig's flopping ears and the gate's rusty wires, and slid it into a long tube. He handed the tube to Tara Smith, PhD, a microbiologist from the University of Iowa. Smith had driven five hours overnight, through the thunderstorm and across plains still scattered with debris from brutal summer floods, hoping to get what she now held: a few samples of pig snot.[1]

Smith is an assistant professor of epidemiology and a well-known science blogger, and deputy director of Iowa's Center for Emerging Infectious Diseases, which investigates illnesses that cross from animals to humans. At the beginning of the summer, she and several of her graduate students had quietly released startling news: seven out of ten pigs on farms in Illinois and Iowa were carrying an odd new strain of MRSA.[2]

Their findings marked the first time that MRSA had been found in a pig in the United States, but a few concerned scientists had seen the development coming. Since 2004, first in Europe and then in Canada, researchers had been finding the previously unknown strain: first in pigs, then in pig farmers, and then in other humans—including health care workers and newborns—who had no connections to pigs at all.

The discoveries represented an odd and troubling development. Smith and her students—including Abby Harper, whose study uncovered the new strain's presence in the United States, and Dr. Mike Male, a swine veterinarian studying for an additional degree—were casting about for more data to help them understand it. That meant finding more pigs to swab, and that was more challenging than it sounded, because few farmers wanted their herds linked to a species-crossing superbug. But after careful negotiation, the Iowa team had gotten access to the Minnesota state fair's swine displays, based on the promise that they would ask each breeder for permission to test their pigs, and would not publicly link any findings back to the farms the pigs came from.

"We know that the first time we looked for this, we found it," Smith said, tucking a handful of sample tubes into a small cooler. "We just don't know enough yet to know what that means."

The discovery of the pig strain of MRSA was an international detective story that started in the Netherlands with a six-month-old.[3]

The toddler was scheduled for heart surgery in July 2004 at Radboud University Medical Center in Nijmegen, a two-thousand-year-old city six miles from the German border. Before the procedure, the staff at Canisius-Wilhelmina Hospital performed a check that is routine in Holland: they swabbed her and cultured the swab to make sure she wasn't colonized with MRSA that could spread to other patients or staff. Dutch hospitals had been doing that since 1988 under a draconian but effective system they called "search and destroy."[4] The system's rationale was simple: the best way to keep MRSA from spreading in a hospital is to never let it in the door. Search and destroy assumes that every patient, and every physician, nurse, and lower-level staff member, is colonized until proven clean.

Patients who were strongly suspected of carrying MRSA—anyone who had been admitted to a hospital in any other country, anyone who had an open, draining wound—went straight into strict isolation on arrival. They and every other patient were checked with multiple swabs, not only in the nose, but in the throat and groin as well. Any positive patient who wanted to stay in the hospital was decolonized with antibiotics and body washes. Until a second set of swabs proved the bug had been dislodged, hospital staff would put on gowns, gloves, and masks before going into the patient's room. If health care workers were exposed to the patients before the patients were found to be carriers, and picked up the bug, they were sent home on paid leave until they were clear.[5]

Search and destroy was costly, irksome, and strict. But it worked. It kept MRSA from multiplying. In 2000, when Netherlands hospitals took samples from patients, they found that only 1 percent of all staph in their hospitals was resistant to methicillin and its chemical cousins.[6] In the United States at the time, the proportion was 50 percent.[7] And because in-hospital levels were so low, MRSA did not spread back to the outside world when patients were discharged. Stringent adherence plus extraordinary microbiological luck—the absence of any United States–style eruption of community-associated staph that created cases outside hospitals—made the Netherlands an almost MRSA-free zone. When newly admitted patients were checked in 2000, only 0.03 percent of them were carrying MRSA that they had acquired outside the hospital. In the United States, the proportion of carriers was 2.6 percent—85 times higher.[8]

So against that backdrop, the discovery of a toddler who was colonized with MRSA was bizarre. "I saw twenty patients colonized in a year, max, and in every case we knew the source," said Dr. Andreas Voss, a professor of clinical microbiology and infection control at Nijmegen, who analyzed the strain. "And I had not seen a MRSA infection in fourteen years. Yet here was this little child, who had not been in a hospital and had not been abroad. It was amazing."[9]

Amazing became unnerving when the hospital postponed the baby's surgery to decolonize her, because the bug simply would not go away. It survived antibiotics, nasal ointment, and antimicrobial washes, stubbornly spreading across the culture gels whenever the nurses swabbed her.[10] Out of frustration, the hospital checked her parents. They were positive as well.

The family were pig farmers; the baby, her parents, and her older sister lived on a small farm near Nijmegen. The entire south of the Netherlands, tucked between Belgium and Germany, is a center for swine raising, a former patchwork of small family properties that is gradually being subsumed by large American-style operations with thousands of animals. The family belonged to a network of farmers who met each month at each others' farms. In November 2004, lacking any other clues, investigators from the hospital and university asked to come to the group's monthly meeting to check the other members for MRSA. By happenstance, the next meeting was at the toddler's family's farm.

The farmers in the network did not spend enough time together, or have intimate enough contact, to pass MRSA to each other. Their only common link was close exposure to pigs. Pigs seemed like an unlikely source of MRSA infection. Swine usually do not carry S. aureus, and the staph species they do harbor—S. sciuri, S. lentus, S. hyicus—do not infect humans. Still, there is a long history of cattle and goats developing mastitis, or teat infections, from S. aureus, and there were occasional mentions in the veterinary literature that those animals had passed that infection to their human minders.[11] Additionally, a researcher in Korea had found traces of MRSA—twelve positives out of 894 samples—in milk from cows with mastitis symptoms.[12]

But the researchers' intuition that the pigs had acquired MRSA and passed it to their farmers was correct. On the first family's farm, one out of thirty pigs was colonized, and an astonishingly high number of the farmers were carrying the bug asymptomatically as well: six out of the twenty-six, or 23 percent, 766 times the rate in the Dutch population. But there was something odd about the strain the pigs and farmers were all carrying: it could not be identified by PFGE, the molecular-fingerprint technique that scientists use to compare isolates. PFGE starts with applying an enzyme that snips the bacterium's genetic material into segments, the first step in achieving the technique's distinctive bar code–like results. But the enzyme, used on MRSA strains in labs all over the world, for some reason did not cut this new strain—so the technique did not deliver a bar code, but instead returned a swath of smears that ran down the gel.[13] Researchers dubbed the new strain NT for "nontypeable."

None of the people carrying the NT strain—the baby, her family, or the other farmers—had been made sick by it, and neither had the one pig the researchers found. So it was possible this cluster was a brief aberration, the latest in staph's near-infinite assortment of evolutionary tricks. However, while the Nijmegen colonizations were being investigated, an actual case of illness—and more colonizations—were uncovered in Weert, a town sixty miles away and even deeper into the pig-farming southeast.[14] In October 2004, a young woman who had just given birth came back to her local hospital with a MRSA breast infection and a high fever. It was the NT strain. The mastitis healed with antibiotics, but she stayed colonized with the bug and could not shed it. The hospital screened her husband and newborn daughter, found they were NT carriers as well, and prescribed a decolonization regimen. Six months later, though, the baby girl developed a bad ear infection, and when she was brought to the hospital, she and her parents were checked again. They were all still carrying the aberrant NT bug. The father was a successful pig farmer, with eight thousand swine on four properties. Researchers from the Dutch National Institute for Public Health and the Environment, the laboratory that screens and classifies every MRSA strain in the country, descended on one of the farms, picked ten of his pigs at random and found eight of them—and three of the farmer's employees—carrying the same NT strain.

The first incident allowed the Dutch to identify the new strain and hypothesize that it passed to humans from pigs. The second demonstrated that it could cause illness in humans, and also pass between humans, because the infant in Weert had never gone near any of her father's pigs. But a third case would prove to them that the new strain was not only odd, but dangerous.[15] In Utrecht—a city in the center of the country, fifty-five miles from Nijmegen, seventy-five from Weert, and well outside the pig-farming areas—a sixty-three-year-old woman who had received a kidney transplant came back to the hospital very ill with endocarditis. Endocarditis is a heart-valve infection that in the pre-antibiotic era killed half of the people who contracted it. The sick woman had NT-MRSA. She had had no contact with pigs, pig breeders, or pig slaughterers. The new strain had spread far enough from its origins to attack her, but there was no obvious explanation as to how.

To solve the mystery of NT-MRSA's emergence, it was important to find out just how widely the pig strain was distributed, in animals and in people. Starting in November 2005, the Dutch checked pigs being brought for slaughter in nine abattoirs around the country. Choosing the slaughterhouses was an uncomplicated method for getting access to pigs from all over the country, because the nine abattoirs handled two-thirds of the pigs slaughtered in the Netherlands every year, and each slaughterhouse bought pigs from a wide range of farms. Over three months, the researchers swabbed 540 pigs: 60 at each abattoir, 10 pigs from most of the farms, with a few small farms contributing a smaller number—fifty-five farms, all told.[16]

The researchers found the new MRSA strain in every slaughterhouse, in 40 percent of the pigs and 80 percent of the farms—though at some farms, only one or two pigs were carriers, compared to four farms where every animal among the ten chosen at the abattoir carried the bug. Microbiological analysis by the national lab uncovered new information about the strain. The samples were diverse, with many small differences—but when the researchers performed MLST, the test that groups strains into sequence-type or ST categories based on small genetic differences, they all fell into one tight group that was designated ST398. The NT strain carried SCCmec type IV, the small, easily transferred genetic segment that identifies community-associated MRSA in the United States, and sometimes SCCmec type V, a newly identified variant. The strain must have been in pigs for a while, because in addition to methicillin and the rest of the beta–lactams, it was also resistant to tetracycline. Unlike in the United States, where tetracycline remained part of the armament against community-associated MRSA, tetracycline was not used against MRSA in Holland because there was no community strain there. But Dutch farmers commonly used the drug for a variety of infections in their pigs—and in the pigs, this new MRSA had picked up tetracycline resistance.

What the microbiological analysis could not reveal was where the strain had come from. It could have been a variant of MRSA that jumped to pigs intact from another species—a cow, or perhaps a human—and then adapted. Or it might have begun as a drug-sensitive staph that lived in pigs undetected and picked up resistance factors piecemeal from bacteria in other animals or in the environment. There was also no way to tell exactly where in pig production the bug was spreading. Pigs in the Netherlands—and in the United States and Canada—are born to sows on breeding farms and then moved at least once, either to a farm that raises them from weaning to slaughtering age, six months, or to an interim nursery farm before the final stop. Because MRSA was found in the pigs' final hours of life, researchers could not determine whether they had acquired the bug at birth, picked it up after weeks or months, or even been infected by other pigs at the same slaughterhouse.

It was clear, however, that NT-MRSA—now renamed MRSA ST398 for the MLST results—was widespread in pigs, and that made investigators worry how far it had gone in humans. Tests on practicing veterinarians in Holland proved the strain was spreading: seven out of 152 vets were harboring it.[17] That equalled 4.6 percent, or 160 times higher than the proportion in the Dutch population. In fact, it was about the same as occurred among Dutch residents who were treated in foreign hospitals. Search and destroy regulations considered travelers such a risk that when they entered Dutch hospitals, they were put into isolation immediately. The appearance of ST398 in humans was such a serious finding that in the summer of 2006 the Netherlands Working Party on Infection Prevention, the national infection-control agency, put pig farmers and swine veterinarians on the same high-risk, "immediately isolate" list.

As part of the search and destroy regulations, the Netherlands maintains national surveillance for MRSA. Hospitals send the first isolate from any newly identified carrier to the national lab. The decision to include vets and farmers among suspect carriers meant that, from the middle of 2006, the lab could begin to build a database of ST398 from patient samples. After a year, they analyzed data from the samples they had received to try to determine any trends. They were amazed to discover that the new strain, unknown just three years earlier, was responsible for 30 percent of all MRSA colonization in the country.[18]

Up until now, the Netherlands' success in restricting the spread of MRSA had relied both on low rates of the disease and also on rigorous infection control. But the mechanisms used to contain MRSA were slow to adjust. At St. Anna Hospital in Geldrop, a town halfway between Nijmegen and Weert, a diabetic patient awaiting surgery for an open ulcer developed a MRSA infection in the wound. It was ST398. Two other diabetics on the ward contracted the same infection. When the hospital widened its search, it found three other patients who were colonized with the pig strain, and five members of the hospital's staff. None of the patients had had any contact with pigs or pig workers. One of the staff members lived in a house on a pig farm's property, but when she described her everyday life for the investigators, it was clear she never came near the pigs.[19]

Search and destroy, which had protected the Dutch against MRSA for twenty years, was breaking down under the wide spread of the new strain. ST398 was colonizing hospital staff. When researchers interviewed health care workers in the southeastern corner of the country, they found that about 3 percent of them had some exposure to pigs in their daily lives.[20] And it was infecting patients also. The number of MRSA carriers identified at Amphia Hospital in Breda, a town in the southwest, went up 300 percent in a year, and the hospital began to have more patients it needed to isolate than it did isolation rooms to put them in.[21]

It was spreading not just within the country, but beyond it. The Netherlands is small, and work rules in the European Union are generous: health care staff and agricultural workers can move freely wherever jobs are available. Agricultural trade is open as well; the Netherlands exports almost 10 million pigs a year for raising or slaughter.[22] So perhaps it should not have been surprising that, in late 2007, three patients connected to the Southern General Hospital in Glasgow, Scotland, developed ST398 infections. Two were infants who were born at the hospital; both had infections of the stumps of their umbilical cords that were discovered by their pediatricians after the babies went home. The third was an adult who developed a wound infection while still in the hospital. All three had identical ST398 strains. No one at the hospital had ever heard of the strain being seen there before, but they checked the databases of their microbiological laboratory just to be sure. Unnervingly, they found that they had already harbored it, and not known it. It had turned up in a swab of the hospital environment taken as part of a research project in December 2006. The discovery reinforced the likelihood that a colonized health worker had transmitted the bug, because the 2006 swab had been taken in a part of the hospital where patients did not go.[23]

Despite that undetected hospital sample, the Glasgow patients marked the first known ST398 strains found anywhere in the United Kingdom, None had been found in pigs there—but then again, no one was looking for it. In Canada, someone was.

* * *

At the Ontario Veterinary College in Guelph, a city west of Toronto best known as the home of Canada's largest brewery, researchers had been monitoring MRSA for at least seven years. But not in pigs—in horses. Dr. J. Scott Weese, a veterinarian and professor of pathobiology, had been interested in the bug since 2000, when it was carried into the college's Veterinary Teaching Hospital by a very sick newborn foal. The infection wasn't recognized until a few weeks later, when a second hospitalized foal came down with the bug. Tracking back, and checking their hunches against clinical records and nasal swabs, the staff realized that several of them must have picked up the infection from the first foal and passed it to the second one.[24] The discovery left Weese and his co-workers and students with tangled emotions. They were intrigued by their scientific discovery, but troubled to be responsible for passing an illness to an already sick animal entrusted to their care.

"There's really close, intensive contact between these animals and their health care providers," Weese said. "The animals flail around and have to be held down and restrained. They're nursed very closely. We have much closer contact than a veterinarian usually does with an animal, or a human health care worker does with a human patient."[25]

The strain Weese's team identified in horses was surprising, as well. It was USA500, called MRSA-5 in Canada. USA500 was a health care strain, one of the first strains to spread in hospitals, initially in Europe, and then in the United States.[26] In the years since it first emerged, though, it had been squeezed out of the hospital environment by other MRSA strains, and was now considered somewhat rare in humans. In horses, it seemed to have found a niche.

In 2007, with growing news of animal MRSA in the Netherlands, Weese, two other Guelph professors, and a graduate student decided to look for MRSA in Canadian pigs. They checked 285 pigs on twenty Ontario farms in what researchers call a "convenience sample"—scientific jargon for "We asked whomever we could think of and took whoever said yes." They found MRSA on almost half the farms, colonizing 25 percent of the pigs and 20 percent of the small number of farm personnel who consented to have their noses swabbed.[27] It was the first time anyone had found the new strain on this side of the Atlantic, strong evidence that the pig strain, and its potential to cause severe human disease, had somehow crossed the ocean. The results also put a new twist on the evolving pig-MRSA story, because both the Canadian pigs and people were carrying not only ST398, but also a recognized human strain, the most common community strain in Canada. It looked to Weese as though MRSA was crossing species in both directions, with ST398 coming from pigs to humans, and humans transmitting the Canadian community strain back to pigs.

The implications of these findings were very significant for the United States, Canada's largest trading partner. Canadian farmers sell more than 9 million live pigs into the United States each year.[28] The state that raises the most pigs—30 percent of the 67 million raised in the United States each year—is Iowa.[29] At the University of Iowa, Professor Tara Smith, the epidemiologist interested in newly emerging diseases, was alert to the possibilities.

Smith and her colleague Mike Male, the veterinarian, were already studying several "production pyramids," large-scale agriculture's term for a closed system of farms in which pigs destined for slaughter spend their lives. Male was doing surveillance for a completely unrelated bacterium, *Lawsonia intracellularis*, which inflames the linings of a pig's intestines and interferes with its digestion—a bad thing when the goal of raising an animal is to get it to put on weight as efficiently as possible.

To test for Lawsonia, Male had to take blood samples from pigs on multiple properties in the multifarm pyramid systems. He had the idea to swab the pigs' noses while he was at it, and gave the swabs to the other Iowa team members. All of them had been intrigued by the pig strain of MRSA, and were wondering when it would arrive in the United States, or if it had already. Male was affable and down-to-earth, necessary qualities for anyone who spends a lot of time wading through reeking, manure-laden muck. He was also an Air Force veteran of the first Gulf War, and possessed a serviceman's practical realism regarding when to ask permission and when not to overexplain. Sampling the swine he cared for would not reveal anything that could harm his clients, and it could help understanding of ST398. "MRSA doesn't bother the pigs, so it doesn't really tend to bother us," he said. "So we don't look under that rock."[30]

With the farms' permission, Male snagged samples from 210 pigs in one pyramid, an operation in eastern Iowa and northern Illinois with 60,000 swine spread across seven farms, and from eighty-nine pigs in a second all-Iowa pyramid of 27,000 swine on two farms. Both sets of farms had sold and replaced their sows just two years earlier. The smaller system bought its breeding stock in Michigan, and the larger one from Minnesota, Illinois, and Canada.

The larger pyramid—the one that had purchased pigs from Canada—

was harboring ST398. Male found the strain in 70 percent of the pigs from its farms, and in nine of fourteen workers who had agreed to be sampled. All of them were colonized, but none showed any signs of illness. No animals or people on the second set of farms showed any sign of the strain.[31]

The results added to the emerging story of ST398, but they also posed fresh questions. The two systems not only bought their sows from different sources, they raised different breeds of pigs. At the colonized farms, the strain was not evenly distributed among all the pigs; it was in all of the youngest ones, and in half of the pigs that had grown to twenty-four weeks, but in only one-third of the two-year-old sows. Among the humans, the workers who performed tasks that should have posed the greatest risk for picking up bacteria, such as taking blood samples, were least likely to be carrying the bug. On the other hand, Male found the bacterium everywhere on the colonized farms. He'd left some culture plates open to the air in the barns for just a few hours, and they'd all grown MRSA by the time he retrieved them. So the most important question might be not why some of the pigs and humans carried ST398, but why all of them did not.

"I think the most important question is, 'Is this strain learning to become a human pathogen?'" Male mused. "And we won't know the answer to that for a while."

In one important respect, the Iowa results were very much like those from the Dutch studies. In both countries, ST398 was resistant to tetracyclines as well as to penicillins and beta-lactams. The resistance pattern was odd; few human community strains exhibit it—possibly because tetracyclines are used against uncomplicated skin infections only for short periods of time and thus do not put much pressure on the bug to develop defenses against them.[32] But tetracyclines are commonly used in swine. Infecting the pigs, ST398 had developed defenses against the drug; crossing to humans, it brought that resistance factor with it. The cross-species leap revived the unsettled debate over the health threat posed to humans by the enormous amounts of antibiotics and antimicrobials used in animals.

Hard numbers for those amounts are contentious and hard to verify. In 2001, the nonprofit Union of Concerned Scientists calculated that 29.5 million pounds of antibiotics and antimicrobials were given to food animals in the United States each year, six times the 4.5 million pounds that humans swallow, apply to their skin, or deploy in antimicrobial soaps. The Animal Health Institute, representatives of large-scale agriculture, put the animal figure much lower in numbers and significance: 17.8 million pounds, compared to 32.2 million used by humans.[33]

Food animals get many drugs for many reasons. They get them for disease treatment. They get them for disease prevention: if one animal in a flock or on a farm is sick or if there is an illness in the vicinity or the farm is thought to be at risk, then all of the well animals get drugs too. Food animals also get antibiotics for "growth promotion," a metabolically mysterious process that has made possible the entire high-volume, low-margin business of industrial-scale farming. Since the 1950s, when two pharma company scientists discovered that feeding chicks the waste products from drug manufacturing made them put on weight much faster, many U.S. farmers have been giving tiny doses of antibiotics to cattle, swine, and poultry.[34] The Union of Concerned Scientists estimates that, of those 29.5 million pounds of antimicrobials given to animals every year, only 2 million of them are actually intended to treat disease. The rest, almost 80 percent of all the antibiotics used in the United States every year, are "nontherapeutic."

The process makes human-medicine experts furious. From their point of view, farmers are routinely practicing antibiotic misuse: giving drugs in the absence of disease, and giving them in such small doses that they kill off only vulnerable bacteria and leave the Darwinian battleground clear for the tough ones. Making it worse, many of the animal drugs are identical, or closely chemically related, to drugs used in humans to combat disease.

For decades, science and agriculture have fought bitterly over whether lavish drug use in animals creates resistant bugs that can and in some cases do attack humans. Proponents of the practice could claim, with narrow scientific correctness, that the case was not proven. There was no single case of drug-resistant illness in a human for which researchers could trace every step in the chain, starting with a particular dose of drug, documenting its administration to a particular animal, proving the rise of a resistant mutant in that animal—and then proving that manure, meat, milk, or eggs from that exact animal had conveyed that drug-altered bug into the system of a particular identifiable human.

But the laboratory and epidemiologic evidence proving linkages between every step in the chain was enormous, stretching across decades of research and hundreds of outbreaks of foodborne illness. Consistently, it has been shown that groups of animals received drugs, and bred resistant bacteria; that groups of people who were exposed to those animals acquired the same resistant bacteria; and finally that groups of people who cared for the animals or ate their meat, milk, or eggs were made ill by them.

In 1976, Dr. Stuart Levy of Tufts University, the bow-tied dean of antibiotic resistance studies in the United States, demonstrated for the first time that resistant organisms arose in Massachusetts chickens that were given feed laced with small amounts of tetracycline, and then spread to humans who tended the chickens. He also proved that both the organisms in the chickens and the identical organisms in the humans learned over time to resist additional drugs that the chickens had never taken.[35] In 1988, partly because of Levy's work, the European Union banned the animal growth-promoter avoparcin because the drug was a close chemical cousin of vancomycin, and contributed to the development of the almost-untreatable human bug called vancomycin-resistant *Enterococcus*, or VRE.[36] The causal linkage was taken for granted by 1999, when a team of researchers at the Minnesota Department of Health established that human illnesses from fluoroquinolone-resistant *Campylobacter jejuni*, a bacterium carried in the guts of chickens, increased after the United States licensed fluoroquinolones for use as poultry growth promoters. The finding led to the cancellation of one new quinolone in 2001, and the FDA's withdrawal of approval for another in 2005.[37-39]

In spite of that research, giving antibiotics for growth promotion and disease prevention persisted in the United States. By the time MRSA ST398 arose in Dutch pigs in 2004, though, it seemed that the focus of the argument had shifted. The contention was no longer that the practice was safe for the humans who took care of the animals, or for those who ate their meat, milk or eggs; instead, it was that it was economically impossible for agriculture to stop. In CAFOs—"concentrated animal feeding operations"—antibiotics were the only way to keep livestock healthy long enough to efficiently put on weight.

There are government rules for what defines a CAFO.[40] The minimums are 1,000 beef cattle or calves or 700 dairy cows; 2,500 large swine or 10,000 swine weighing less than 55 pounds; 55,000 turkeys, 5,000 to 30,000 ducks, or from 30,000 to 125,000 chickens. CAFOs can often be far larger; in a 2008 report, the Government Accountability Office described farms with 800,000 hogs and 2 million chickens.[41] Amazingly, no one really knows how many CAFOs there are, because no government agency is charged with keeping track of them. The GAO estimates that there were almost 12,000 in 2002, three times more than there were twenty years earlier.

From a microbiological standpoint, the problem with CAFOs is not just the drugs given to the animals, or the vast numbers of animals, which increase the chances of a resistant germ evolving, or the miserable crowding that creates a perfect setting for passing resistant bacteria from one animal to another. It is also what those animals leave behind: more than 300 million tons of manure a year, twice as much as comes from all the humans in the United States.[42] A single large CAFO can produce more than 1.6 million tons of manure in a year, more than the entire city of Philadelphia. On a small-scale farm, the manure would be sprayed on cropland, but there isn't a lot of cropland near CAFOs. Instead, there are other CAFOs, clustered in tightly defined areas: northern Iowa and eastern North Carolina for pigs, western Wisconsin for cows, and chickens in northern Georgia, northern Alabama, western Kentucky, and western Arkansas.[43] With nowhere to spray it, the manure is stored on farms in enormous lagoons. Some gut bacteria survive in manure, and so there are bacteria in the lagoons. Some of them may be resistant bacteria, carrying resistance genes that are available for other bacteria to acquire. If any antibiotics are being used on the farm, there will be antibiotic residues in the manure as well, putting additional evolutionary pressure to develop resistance on whatever bacteria are present.[44]

All of this is bad enough, but the manure in lagoons does not stay there. Huge amounts escape when lagoons collapse; a lagoon break in 1999 dumped 20 million gallons of hog waste into North Carolina's New River.[45] There are more subtle leaks as well. In Illinois, researchers investigating giant hog farms' influence on the local environment found the genes for tetracycline resistance in groundwater near CAFOs.[46] In Minnesota, scientists tested the safety of using CAFO waste as fertilizer by growing plants in manure contaminated with antibiotic residues, and found the antibiotics had been taken up into the leaves of lettuce and the flesh of potatoes and corn.[47] When Texas researchers looked into whether resistant bacteria were escaping CAFOs, they found MRSA 150 meters away from a Midwestern hog farm, carried by the wind.[48] In Maryland and Delaware, home to enormous indoor chicken farms, other researchers found resistant bacteria carried long distances by flies.[49] Bacteria also walk out of CAFOs on people. In 2007, Johns Hopkins researchers tested a group of workers from Delaware and Maryland chicken farms and found that half of them harbored multidrug-resistant E. coli. The researchers were especially disturbed to find that most of the workers brought their dust-, manure-, and presumably bacteria-laden clothes home to wash.[50]

CAFOs, in other words, have the potential to be enormous, unpredictable, leaky Petri dishes for breeding resistant bacteria and for dispersing those bacteria into the surrounding environment. This was the setting in which MRSA ST398 first appeared, and is the reason the first researchers to recognize it were made so nervous by its arrival. Their only comfort, if there was one, was that the most lethal bacteria that had become resistant on farms had been ones that caused foodborne illnesses, from organisms that live in animals' guts and contaminate their meat at slaughter. ST398, though, had passed to farmers, their families, and other unexplained victims by colonization, in a chain that presumably began when someone somewhere came near the nose of a pig. There was as yet no suggestion that this unpredictably fatal new MRSA could behave like a foodborne disease.

But researchers' fears for ST398 would soon be realized.

In the Netherlands, the emergence and rapid ascent of ST398, and the ease with which it subverted two decades of stringent infection control, was a serious shock. Researchers there almost immediately made the intuitive leap: if a novel strain of MRSA was spreading in food animals, were food products also an infection risk?

As a hypothesis, it seemed a stretch. The drug-resistant bugs already passed to humans by animals—salmonella, campylobacter, and E. coli—not only come from the gut, they attack there too, when humans swallow them. MRSA was racking up a long list of body sites it could invade—skin, muscle, lung tissue, bone—but all of those infections began with MRSA getting access through the skin. Some other strains of staph can cause ferocious foodborne illness by producing toxins after they are swallowed, however, and there was one odd case in the medical literature that suggested what a lethal combination MRSA and food could be. In the winter of 1992, when the bacterium was still considered only a health care pathogen, twenty-one patients in Rotterdam's University Hospital developed serious blood and wound infections from the bug despite stringent search and destroy efforts. Five died. After six months of investigation, the hospital traced the outbreak to a member of the hospital's dietary staff who was colonized with MRSA in his throat. He had never gone into the patients' rooms or had any direct contact with them, but he had exhaled the resistant bug onto their food.[51]

The Dutch researchers who had first uncovered ST398 began looking for the strain in meat, and soon demonstrated that their hypothesis was not as far-fetched as it seemed. After buying and testing seventy-nine samples of pork and beef from thirty-one butcher shops and supermarkets, they found the NT strain on one piece of pork. In December 2007, they announced: "MRSA has made its way into the food chain."[52]

In Canada, Scott Weese had been thinking along the same lines. He executed a similar survey, recruiting colleagues to buy 212 samples of butchered and ground pork in stores in four of the ten Canadian provinces. They found ST398 in nineteen of them.[53] Weese presented the findings at a scientific conference, sliding them in almost unnoticed, in one of the last sessions of the week, at the end of a discussion of the emergence of ST398.

It was too early, he said afterward, to be able to say just how much of a risk the new strain posed. The possibility that MRSA-contaminated meat could cause severe disease was probably slight; many of the victims in that 1992 Rotterdam outbreak had been cancer patients whose immune systems had been undermined by chemotherapy. The most likely risk seemed to be from people handling raw pork with less care than they should. From years of warnings about raw chicken, everyday cooks had learned to wash their hands, segregate cutting boards, and keep blood and juices away from produce and other raw foods. "But people don't tend to handle pork like it is biohazardous," he said.[54] If they handled the raw meat carelessly and then touched their eyes or noses, they could become colonized with the new strain.

The CDC was not so concerned. After Weese found ST398 in North America, the agency's then-director Dr. Julie Gerberding, was summoned before Congress to discuss the possible threat. "Although the finding of MRSA in retail meats suggest a possible role for foodborne transmission, if such transmission occurs, it likely accounts for a very small proportion of human infections in the United States," she said.[55]

But for Weese, and for Tara Smith in Iowa, infections were not the biggest issue. The most unnerving aspect of ST398 was its rapid and unmeasured spread. The strain had crossed from Europe to Canada and from Canada to the United States, and it was almost certainly crossing the U.S. now too. Every year, millions of pigs are raised in one state and fattened to slaughter weight in another. Fewer than one-third end their lives on the farm where they were born.[56]

The pig strain was crossing into the country via other routes as well. In the summer of 2008, New York City researchers found a drug-sensitive version of ST398—with the same distinctive evolutionary markers, but without resistance to methicillin and the beta-lactam drugs—in residents of Manhattan and in their relatives in the Dominican Republic towns from which they had emigrated. So many people in those communities flew between them so frequently that researchers called the close connection between the two locations an "air bridge," though no one could say in which direction the bacterium had crossed first. They were plainly concerned that the bug would pick up that missing resistance, and that MRSA ST398 would find a home in New York City.[57]

The Manhattan and Caribbean residents were colonized with no signs of infection, like the pigs and the pig farmers that Tara Smith and her co-workers had identified. Sitting in her lab in Coralville—not far from a lunch counter that advertises "the biggest and best pork tenderloin in Iowa," a sandwich that resembles an old vinyl record accidentally inserted into a bun—Smith mused on the effects of ST398 moving through crowded animals on farms and crowded humans in cities, encountering antibiotics, entering hospitals, adapting toward some unpredictable, potentially dramatic change.

"It's never a good thing to have a high percentage of a population colonized with MRSA, whether it's animals or humans, the hospital strains or the community strains or this one," she said. "But what that is going to mean, in this case, we just don't know."

CHAPTER 11

SOAP, THE WONDER CURE

Despite years of pressure from activists, and several attempts at national legislation, there was no government regulation to address the use of antibiotics in farm animals and the resistant bacteria that resulted. The situation was surprisingly similar in another major breeding ground for resistant bugs: hospitals. There were few universal standards for controlling and preventing hospital-acquired infections, and none for hospital MRSA; and surveillance for the bug was purely voluntary. Yet in cities scattered across the United States, hospitals were undertaking to tackle MRSA, sometimes on their own, sometimes with the help of grassroots organizations or under pressure from activists. No two efforts were exactly alike, because they were tuned to the local culture, and to the collective personality of the institutions that developed them. Taken together, they demonstrated that controlling hospital MRSA was not an unreachable goal.

It was July 2008 in North Carolina, and it was steamy. Humidity hung in the air, hazed the sunlight, and beaded like rain on the tinted-glass windows of Charlotte's Presbyterian Hospital. On days like this, when the air was so thick it felt solid when you breathed, patients with chronic diseases struggled. The asthma clinic was slammed, the emergency room was filling, and the dialysis unit on the fourth floor of the hospital's oldest building was bustling with patients being run in and out in wheelchairs and on gurneys.

Tucked in a corner, registered nurse Patti Deltry watched the traffic. The dialysis unit was almost always busy. Charlotte is a growing city, and as its population increases, its rates of diabetes and kidney failure, the ills of the Deep South "Stroke Belt," have been rising too. The eight beds in the small dialysis unit were almost always full of patients waiting listlessly through the four hours it took to clean their blood of toxins as it circulated through pumps and tubes. The patients were from the intensive care unit and the critical care service, people whose borderline kidneys had shut down without warning. They were medically fragile and needed expert care; nurses could not work in the unit until they had two years of experience, and some of the senior ones wore twenty-five-year service pins.

But even very experienced nurses can sometimes be distracted, and that was why Deltry was loitering by the biohazard bins and observing the work flow with unswerving attention. She watched as a nurse clad in the unit's distinctive blue scrubs snapped on her gloves to start caring for a new arrival. The nurse stepped toward the bed, but stopped first to page through the phonebook-thick chart that had accompanied the patient from elsewhere in the hospital. Quietly, so as not to attract attention, Deltry stepped forward and spoke softly to the nurse. The nurse blushed, nodded, stripped off her gloves and headed to the sink to wash her hands, before gloving again and going straight to the bedside.

"She hadn't touched the patient yet or gone from one patient to another," Deltry said a few minutes later, walking upstairs to another unit where patients recover from coronary-artery bypass surgery, "but it's our policy here that no one walks around with gloves on, because you could get distracted and not do the right thing. And everyone wants to do the right thing."

On the fifth floor, Deltry took up position again, this time with her back against a narrow strip of wall between the doors to two patient rooms. From where she stood, she could see down two hallways, and behind her back she could hear the automated buzz of the hands-free sanitizer machines installed in the rooms behind her. For the next hour, she would observe the health care staff on ward 5C, intervening when she thought education would help and making notes for her reports. Later in the day, she would walk to a third ward, and then a fourth. The next day, she would go to another one of the four hospitals in the Charlotte region that belong to Novant Health, Presbyterian Hospital's corporate parent, and she would perform the routine again, watching, waiting, and making herself obvious.

Novant Health Inc. is a large health care system, encompassing nine hospitals and seven other institutions, 2,650 beds and 24,600 employees. The company offers highly technological medicine, and over the years has tried a range of sophisticated, sensitive programs to keep hospital-acquired MRSA under control. Nothing succeeded as well as Deltry and another nurse colleague observing—and allowing themselves to be seen observing—simply to make sure that employees kept their hands clean.[1]

It took a long time for hospitals to attempt to contain MRSA's spread. From the bug's earliest U.S. appearance in the late 1960s, through the 1990s, hospital MRSA gained ground so relentlessly that many physicians claimed nothing could be done. As late as 1998, with MRSA rates in hospitals surpassing 50 percent of all staph infections, most hospitals did not have any kind of control program.[2] About two-thirds had surveillance programs that used cultures to identify infections occurring in their institutions; that was how they knew how much MRSA was in the building. But hospitals were not putting the information to any preventative use.[3]

By the late 1990s, hospitals were routinely practicing the basic infection-control techniques that had been reinforced by the arrival of AIDS in the 1980s, emphasizing the use of gloves to avoid contact with body fluids. These techniques had no impact on MRSA. Institutions tried restricting the use of antibiotics, to keep from growing their own resistant strains, or rotating patients' antibiotic regimens, to frustrate emerging resistance. Neither effort curbed MRSA's expansion.[4] Isolating patients who were known to be infected likewise had no significant impact on keeping MRSA under control.[5] Some researchers even argued in reputable medical journals that attempting to prevent individual MRSA infections was doomed to failure, and was a distraction from the hard work of shutting down in-hospital outbreaks when they inevitably began.[6]

The pessimists were laboring under an overload of information—more than twenty thousand medical journal papers had been published since methicillin resistance was identified in 1961—that somehow had not translated into any actions that worked.[7] Many of those papers did, however, explain in depressing detail how formidable a foe MRSA could be. In hospitals, MRSA was everywhere. It lingered on floors, bed rails, and tables; on blood pressure cuffs and thermometers; and even on soft surfaces such as bed linens, doctors' white coats, and nurses' protective gowns.[8] Anywhere the bacterium survived, it could be picked up by a health care worker and transferred to a patient. It seemed impossible that a hospital could ever be clean enough that it would not pose a risk to patients inside it. That was especially true given that—as the 1980 Harborview hospital outbreak had shown, and multiple surveys conducted since then in ICUs and emergency rooms and on hospital wards had proven time and again—no more than half of health care workers consistently washed their hands.[9-11] So it seemed improbable that MRSA in a hospital could ever be curbed.

Outside the United States, though, MRSA was being controlled. In the Netherlands, search and destroy kept MRSA incidence at a tiny fraction of U.S. levels. In Denmark, where the national health care system can exert close control over hospital practices and drug prescribing, MRSA had been reduced from 18 percent of all staph isolates in the late 1960s to 1 percent in the late 1970s, where it stayed for more than a decade.[12] Those countries were not perfectly analogous to the United States, because they had centralized health care and medical records, homogeneous populations, and a lower prevalence of hospital-acquired MRSA than the United States had experienced in decades. But they demonstrated that attention and will could change the course of a national epidemic.

Dr. Barry M. Farr, a hospital epidemiologist at the University of Virginia, believed the United States had both the obligation and the ability to reverse its national epidemic, though he was not convinced the country possessed the will. Farr was finishing medical school at Washington University in St. Louis when the University of Virginia Medical Center checked in its first case of MRSA on March 13, 1978, a man with endocarditis that released septic emboli into his circulatory system. Between March and September, an outbreak flared. The strain went from the first patient to a second, who gave it to three others who shared his room. They brought it to the intensive care unit, where it leapfrogged from bed to bed. Thirty patients were found to be colonized in the first six months; sixteen developed infections, and three died. The strain spread further throughout the medical center. Within about fifteen months, it was causing 24 percent of all bloodstream infections and 38 percent of all surgical infections there.[13]

Farr arrived at Virginia as a newly minted medical resident in the summer of 1978, in time to witness the epidemic spreading through the burn unit and ICUs where patients were most vulnerable. The hospital had scrutinized every potential reservoir of the germ. It isolated the infected, screened more than five hundred employees for MRSA carriage, and scoured the buildings for evidence of environmental contamination. When those efforts had no impact, administrators turned to the one remaining potential source, uninfected patients who showed no symptoms, but who had unknowingly brought the bug into the hospital. Starting in December 1980, the hospital began looking for MRSA not only in patients with obvious infections, but in anyone who was likely to be colonized, including patients who had been in the hospital before and those newly admitted from other health care facilities where MRSA might be prevalent. There was no attempt to decolonize them; instead, the colonized patients were housed together under isolation, and the health care workers assigned to them were put under strict rules for hand washing.[14]

Virginia Medical Center's program resembled the Danish and Dutch campaigns, but was not based on them; in 1980, the details of national search and destroy efforts were not widely published.[15] In the hospital's view, the plan was a desperate attempt against overwhelming odds, since no one was known to have successfully controlled MRSA's in-hospital spread. It represented a paradigmatic shift, away from the long-standing technique of passive detection—essentially, waiting for the clinical microbiology lab to report that an already recognized infection was caused by MRSA—to active surveillance: testing everyone who might be harboring the bug before it could cause infection in them or others. To the hospital's surprise, the plan worked. In the month that surveillance and testing started, there were twenty new cases of MRSA. The count dropped to ten in January, then to seven in March. After that, month by month, the line on the chart sank toward zero. Despite repeated reintroductions, rates of MRSA at University of Virginia Medical Center continued to stay low for a decade.

The experience made Dr. Farr an evangelist for the new technique, which came to be called active surveillance culturing or active detection and isolation (ADI). Active surveillance not only kept Virginia's rate of hospital-acquired MRSA low, it exposed and short-circuited an outbreak in the neonatal intensive care unit and in an influx of colonized patients transferred from nursing homes.[16,17] Using the data that the surveillance cultures provided, Farr was able to show that doing tests only after infected patients were found would identify just 15 percent of the MRSA present in the hospital, leaving a vast undetected reservoir from which it might be spread.[18]

In 2000, Farr founded the Problem Pathogen Partnership, a network of hospitals in Virginia and North Carolina that committed to practicing ADI using mutually agreed-upon protocols that identified patients at high risk of colonization and dictated how often during their hospital stay they should be checked.[19] In 2002, as president of the Society for Healthcare Epidemiology of America, Farr convinced the organization to issue a set of guidelines that recommended screening high-risk patients as a crucial first step in patient care, saying it was just as critical for MRSA control as hand washing, conservative prescribing of antibiotics, and decolonization.[20] By 2008, five years after the guidelines were published, Farr estimated that more than 160 scientific papers had endorsed active detection and isolation for combating hospital spread.

Novant Health's MRSA control campaign, the program that employed Patti Deltry, was prompted by a tragedy—and by the company's decision to face up to it. A baby boy was born in 2004 at a small hospital in its network, many weeks premature, weighing less than two pounds. He was transferred to a larger hospital, and improved for several days, but he developed an infection and died when he was twenty-eight days old. The infection was MRSA. Shortly afterward, a second child in the same neonatal ICU died of MRSA too. When the hospital checked the rest of the babies in that nursery, they found eleven of them colonized as well.

Paul Wiles, the president of Novant, had worked for years at the hospital where the two babies died, and he took their deaths personally. At the time of the babies' deaths, Novant already had a MRSA-prevention program; in fact, it belonged to Farr's Problem Pathogen Partnership. Taking its cue from Farr, the organization had focused its attention on identifying MRSA carriers among ICU patients, former patients coming back to the hospital, and patients newly admitted from nursing homes. Though it had reduced MRSA rates, the numbers weren't low enough to protect the newborns who died. A survey shortly afterward showed that in 2005 there were 234 cases of hospital-acquired MRSA infection across the Novant network. Another survey indicated why. Only 45 percent of the hospitals' staff regularly washed their hands.[21]

The number was appalling but unexceptional. It matched low rates of hand washing found by many other surveys conducted around the United States over the past few decades. Those surveys had been done by researchers scoring their own hospitals, who might have been tempted to downplay their results, or by outsiders who would publish the results in medical journals months afterward. Wiles was no outsider and could not dismiss Novant's findings: he had to present the results to the hospitals' board of trustees, face-to-face.

"These are the community leaders, and the heads of major national banks," he said. "And their jaws just dropped. 'Your doctors and nurses aren't washing their hands? You're kidding me, right?' And I couldn't defend it."

Wiles brought the problem to the organization's executive team, which agreed that "hand hygiene compliance" would be its top priority. They would commit Novant to reaching the unheard-of hand-washing rate of 90 percent of opportunities—double their current level—in three years. To demonstrate their commitment to the rest of the staff, they linked their compensation packages to it. No executive bonuses would be paid if they were not at 90 percent by the end of 2007.

The campaign would be combating the inertia of established routines and the perceived stress of adding tasks to already-packed days. Washing his or her hands properly before and after every patient can add more than an hour to a nurse's average day.[22] But the effort was also going up against something so deeply ingrained in medical personnel that it is taken for granted and almost never spoken aloud: the impossibility of believing that they could harm the patients they have sworn to care for.

"I know what kills someone. It's a ruptured abdominal aortic aneurysm, or a gunshot wound to the vena cava," said Dr. Thomas N. Zweng, the chief medical officer of Novant's Charlotte division and a general surgeon for almost twenty years, who was also one of the leaders of the hand-washing effort. Putting himself in the shoes of his fellow surgeons, he continued: "I can fix that, because I trained long and hard to learn how. But for me to believe that these instruments that I worked so hard to perfect, these hands that have absolutely saved lives, can kill someone? It's just unfathomable."

To make clear that the whole organization needed to be committed to the goal, Novant called in representatives of the whole organization: administrative staff, nurses, and physicians from multiple departments, as well as the infection-control professionals, the department dedicated to clinical improvement, and marketing. They developed a program, essentially a massive internal PR campaign, that was complex in its execution, but simple in concept. Novant would impress on its employees that hand washing was nonnegotiable. It would watch them, to see that they complied; it would let them know they were being watched; and it would follow up if they failed with education, and with consequences.

Starting in 2005, the organization blanketed its hospitals with messages, plastered on billboards over the parking lots, posters in the hallways, animated screensavers on the computer monitors, decals on bathroom mirrors, and signs on the back side of bathroom doors. Some were intended to make people notice and laugh. One sign over the sinks said, "Send germs where they belong," alongside an arrow pointing down the drain. Others, though, were confrontational, such as an image of a boy patient in a bed, with the caption "You could kill me with your bare hands." Novant put its entire staff, including office workers who never encountered patients, through mandatory education, and installed alcohol gel dispensers everywhere, including administrative offices and conference rooms where patients never went. It also hired two hand-hygiene observers—Patti Deltry was one, though not at first—to monitor and record compliance. Each observer compiled a report at the end of the day that was emailed to the administrative leadership and to the manager of any nurse or lower-level staff who had transgressed. Physicians got a polite but firmly worded letter, signed by Chief Medical Officer Zweng.

The efforts did not go smoothly at first. "There was a lot of angst, a lot of pushback—people wanting to explain themselves, or give you the reasons why that didn't work for them at that particular time," said Dr. Jim Lederer, Jr., Novant's corporate medical director for clinical improvement. Nurses and mid-level staff complained of being spied on, and so many doctors dragged their feet that Zweng threatened to post the names of transgressors on the walls of the doctors' lounge. At the same time, the hospitals' administration struggled with how hard to push physicians, realizing they could relocate their practices and income to another health care organization if they felt aggrieved.

The first two hand-washing observers didn't last. Both were licensed practical nurses, too low on the hospital's scale of status to make an impact speaking to more-credentialed staff. "They were muscled out," Zweng admitted. Deltry and another registered nurse were the replacements. Deltry had been an operating-room nurse. Though polite and soft-spoken, she was used to the famously abrasive personalities of surgeons, and could hold her ground.

Slowly, the initiative took hold. In the initial survey in 2005, rates of MRSA infection had been 0.67 for every 1,000 days patients spent in the hospital. That dropped to 0.5 per 1,000 days in 2006 and 0.42 per 1,000 in 2007. By mid-2008, the MRSA rate was down to 0.3 infections per 1,000 days. There were 193 MRSA cases in 2006, 170 in 2007, and 123 in 2008, at a time when community MRSA incidence was rising across the country and increasing the chance that incoming patients would be colonized. From its internal accounting, Novant estimated that each hospital-acquired infection cost an extra $10,000 to treat. Based on that math, it saved more than $1 million over three years.

In January 2008, the organization zoomed past its once lofty goal of 90 percent hand-washing, achieving 99 percent in hospitals on one side of the state and 98 percent in the other. Enthusiasm had overtaken reluctance. Still, the campaign could not declare victory and stop. Novant, like any major business, has constant slow turnover, about 15 percent of its staff or roughly three thousand people per year. So there would always be new people to be trained, and trained people to be reminded.

"I think of it this way," Wiles said. "You know how, when your kid goes to kindergarten and learns about safety things, they can't wait to tell you about it? The teacher tells them, 'Always wear your seatbelt,' and then they come home and they nag you, 'Daddy, did you buckle your seatbelt?' every single time you go somewhere? That's what we're doing here. We're the six-year-old in the car."

It would be hard to find a U.S. hospital that would say it did not want to control MRSA. Yet few seemed willing to take measures as aggressive as Novant's close scrutiny or Virginia Medical Center's active surveillance and testing. In 2002, 28 percent of hospitals surveyed nationwide said they were using active surveillance—a big jump from the zero hospitals that had practiced the technique before Farr publicized it, but still a small proportion.[23] The technique was dogged by papers in major journals that claimed that it was costly to set up, that it focused too narrowly on MRSA to the exclusion of other hospital pathogens, and that it didn't prevent infections any better than standard precautions.[24-26] Crucially for a medical culture that increasingly focuses on justifying expenditures through evidence, several comprehensive reviews concluded that the many studies of detection and isolation differed so much from each other that they could not be used to build a collective case for the practice.[27,28]

Yet at every major infectious-disease meeting, dozens of small presentations demonstrated that some hospitals were willing to tackle the difficult technique, and succeeding with it. The University of Colorado Hospitals even said it was cost-effective.[29] Memorial Hermann Hospital in Houston reported that it reduced rates of ventilator-associated pneumonia.[30] University of Maryland's medical school showed that it reduced ICU bloodstream infections.[31] Those were very large hospitals, but small institutions also endorsed it: Aspirus Wausau Hospital in Wisconsin cut its number of colonized patients by two-thirds in a year.[32] Some of the most prestigious hospitals agreed, including the Brigham and Women's Hospital, one of the teaching institutions of Harvard Medical School. It successively deployed four different interventions in its ICU: barrier precautions when central venous catheters were being placed; alcohol hand-rubs; an aggressive campaign to promote hand hygiene; and screening patients for MRSA colonization when they were brought to the ICU and once a week thereafter. The first three had no significant impact, but when active surveillance was introduced, infection rates plunged. The impact spread beyond the ICU, with declines in infections throughout the hospital.[33]

Even with such impressive results, there was no health-industry consensus or government guidance over whether to use active surveillance, and in their absence disagreements grew remarkably bitter. In medical journals, researchers traded accusations of misinterpretation and outright misrepresentation with barely masked hostility. "We thought that [the authors] would appreciate hearing about some of their own errors and misrepresentations (all of which cannot be addressed in a letter)," one publication began.[34]

When the government finally did weigh in, the debate only got murkier. In 2006, the CDC published its own guidelines for preventing MRSA and other multidrug-resistant organisms in hospitals. The guidelines were written by the CDC's Healthcare Infection Control Practices Advisory Committee of outside experts, called HICPAC. They had been ten years in the making, and drafts had been circulating and causing disagreements in the infection-control world for at least two years.

The earlier guidelines written for hospitals by the Society for Healthcare Epidemiology of America (or SHEA), the professional organization that Farr had headed, put active-surveillance cultures at the top of a list of things to do for MRSA and hospital-associated infection control. The CDC guidelines did not—but they did not tell hospitals not to do active surveillance, either.[35] Instead, they trod a rigorously observed but unhelpful middle path. Painstakingly reviewing all the evidence for and against active surveillance cultures, the new guidelines hedged: "Their use should be considered in some settings." It was a second-rank recommendation, buried among dozens of others, which left hospitals who were seeking to do better without any help setting priorities. In fact, the CDC guidelines gave so little clear direction that in 2008 the Government Accountability Office publicly scolded the CDC and other agen-

166

cies in the Department of Health and Human Services for not providing better leadership.[36] In the HICPAC guidance alone, the GAO pointed out, the CDC had made eighty recommendations. And that was just in one publication. Since the 1980s, the CDC had issued guidelines about 1,198 specific infection-control practices, and more than 500 of them were "strongly recommended."

Meanwhile, public awareness that it was possible to control MRSA in hospitals was streaking ahead of government deliberation. Consumer and health care organizations endorsed active surveillance, from the advocacy Committee to Reduce Infection Deaths, founded by former New York lieutenant governor Betsy McCaughey, to the Institute for Healthcare Improvement, a nonprofit that focuses on identifying hospitals' successes in reducing harm to patients and helping other hospitals learn from them.[37] Riding the wave of enthusiasm, some hospitals went further than the original protocol of swabbing and culturing patients who seemed at high risk of colonization. They began testing every patient that came in the door.

The model for this aggressive practice was Evanston Northwestern Healthcare, a three-hospital system north of Chicago. Researchers there instituted a twelve-month program of active surveillance followed by isolation and decolonization among patients admitted to their ICUs. Though the program identified MRSA colonization, it did not reduce the number of MRSA infections occurring in the ICU. Then they expanded the measures, testing every patient admitted to the hospital, isolating and decolonizing everyone who was positive, essentially re-creating Dutch search and destroy. With this method, they cut the number of subsequent MRSA infections in the hospital by more than half.

The hospital concluded that relying only on cultures of visibly infected patients in the ICU would have identified only 18 percent of those who were colonized. Even active surveillance just in the ICU would have identified only 33 percent.[38]

Universal screening was effective, but it was logistically complex and required significant investments: time to train staff in new procedures and money for lab personnel and technology to handle the much larger number of tests. Evanston-Northwestern calculated its upfront costs at about $650,000.[39] The single biggest endorser of the practice was the Veterans Health Administration, which in 2007 ordered all of its hospitals to screen and isolate, but not decolonize, patients who were being admitted to any unit other than the psych wards.[40]

But many physicians and hospital administrators continued to ask for more randomized trials, contending that the case for active surveillance, let alone universal screening, had not yet been made. The public, though, was restless. Outrage over hospital-acquired MRSA was growing. Citizen activists were emerging who were unwilling to wait for any more evidence to be gathered. They were MRSA victims themselves, or had MRSA in their families. They insisted on the need to identify and count every instance of MRSA in hospitals. And when hospitals did not move fast enough, some of the activists—such as MRSA victim Jeanine Thomas of Chicago—were willing to find lawmakers who would force hospitals to comply.

The winter of 2000 was a bad one even by Chicago standards: heavy snow, gale-force winds, sheets of ice. The winter was particularly dismal for Jeanine Thomas, a forty-five-year-old marketing executive and antiques dealer who lives in the city's affluent western suburbs. On November 17, she slipped on an icy patch, fell, and fractured her ankle, shattering the joint so completely that putting it back together required a steel plate torqued to the remaining bone with screws. After two days in the hospital, she went home with crutches, ruing her clumsiness and hoping the foot would heal by the start of tennis season. She was a highly ranked amateur player, and she looked forward every year to shaking off the imprisonment of winter and exercising again.[41]

A day after her surgery, her leg began to ache. Two days later, the pain was excruciating. Her foot felt swollen inside the cast, and she could feel the edge pressing into her leg. She called her surgeon. He sent her to the emergency room. In the ER, a resident cut the cast open, and Thomas got a glimpse of her ankle. It was dark, almost black, and the incision was oozing pus.

She needed surgery quickly, the doctors told her, but it was late in the day and she already had eaten, not knowing she would need to go under anesthesia. Surgery would have to wait until the next day, a Saturday. Thomas agreed. It was the last thing she remembered until Monday, when she woke, still feverish, to a nurse telling her that the collection of steel plate and screws had been infected down into the bone.

Thomas's fever soared past 104, and she floated in and out of awareness, once feeling her consciousness rise up from her bed and look back at her body. "I remember," she said afterward, "I woke up, because a nurse was screaming, 'Page the doctor. Give me—' and then I went under again." She was carted back to the operating room for debridements to clean out the dead flesh, excavations that carved cartilage out of the joint and laid her leg open to the bone. She was enduring the ordeal largely alone. Chicago got forty-two inches of snow that December, as much in one month as it usually gets in a season, and her friends couldn't get to the hospital to sit with her and question caregivers on her behalf. Once, she emerged from a stupor in time to catch physicians discussing whether to amputate her foot. After three weeks in the hospital, she went home with a prescription for the oral antibiotic Levaquin, to which she reacted badly, with vomiting and diarrhea. She was in bed for four months that winter—until the start of spring, which would never again mark tennis season for her.

No one had explained why her infection was so intractable. She discovered the cause when she was making a follow-up visit to one of her physicians, an infectious-diseases consultant. Flipping through her chart, he noted, "Osteo."

The word was unfamiliar to Thomas. "Osteoarthritis?" she asked.

"Osteomyelitis," he corrected. "You have MRSA osteomyelitis."

It was the first time she had heard the bacterial acronym, and it meant nothing to her. In 2001, there was little public knowledge of MRSA. Thomas prowled the Internet for information, and her sister, a registered nurse, brought her journal articles to read. She struggled with the aftermath of her infection. The hardware in her leg was removed because infection was lurking in its niches, leaving her with a floppy, unstable joint in which bone scraped painfully on bone. MRSA moved into her sinuses, causing repeat infections so prolonged that the amount of antibiotics she took bred her own resistant strains.

In 2002, she read a *Chicago Tribune* investigation of infections occurring in hospitals, and realized that there were other patients like her.[42] In 2003, she heard that the Illinois legislature was considering a bill that would require hospitals to publicly report infection rates, and recognized a cause she could join. She wrote legislators, called, and made lobbying visits, and was enormously encouraged when the Illinois Hospital Report Card Act passed in midyear. It was the first such law in the country.

With the help of her state senator, Christine Radogno, she got placed on the advisory board overseeing the new law when it went into force in 2004. Thomas was the only consumer representative in a group of public health experts and hospital attorneys. Feeling a new sense of power, she started an organization, the MRSA Survivors' Network, to encourage other patients to seek support and tell their stories.

Somewhat to her own surprise, Thomas had become an advocate. From her perch on the advisory board, she kept an eye on the precedent-setting legislation she had supported. Two years after its passage, she realized that its force was being diluted by quiet regulatory rewrites and that the changes it mandated were not being made. The data gathering that had been supposed to start in January 2004, exposing Illinois hospitals' infection rates, had not even begun.

"I tried to bring attention around to MRSA, asking that it be included in the infections that would be reported, and everyone else on the advisory board voted against me," Thomas said. "It was like a light bulb going off. I thought, *If these officials don't want these rates reported, they're not going to do anything about this disease.*"

She went back to Senator Radogno. Together they crafted another bill that would force the state to target MRSA specifically, invoking the SHEA guidelines that call for surveillance, testing, and reporting of any "at risk" patient, anyone admitted to an ICU or from a nursing home, or with a prior history of MRSA infection. They introduced the bill in January 2006, and watched it stall in committee from health-industry opposition and die.

Frustrated but not cowed, Thomas began to seek out other MRSA advocates. Around the country, she discovered, a citizens' movement was growing. People like her had turned their outrage over hospital infection—and sometimes their grief over a loved one's death—into passionate activism. In Maryland, Michael Bennett founded the Coalition for Patients' Rights after his eighty-eight-year-old father died from the combined effects of six different hospital infections. In California, Carole Moss's Nile's Project memorialized the MRSA death of her fifteen-year-old son Nile. Victoria Nahum in Georgia founded the Safe Care Campaign with her husband Armando after three hospital-caused infections occurred in her family in ten months, in her father-in-law, herself, and in her son Josh, who died.

Thomas and Radogno tried again with their bill in 2007. Public opinion was shifting; they won the unlikely support of the Illinois Hospital Association, Illinois Nurses Association, and Illinois State Medical Society. The legislation passed. The MRSA Screening and Reporting Act was signed into law August 20, 2007, and became effective immediately. The act called for two things: first, a baseline report of MRSA rates in hospitals, drawn from whatever existing statistics the state government could gather; and after that, real-time information on new cases reported by the hospitals themselves. In late 2008, the Illinois Department of Public Health released the baseline data, a count of hospital-related MRSA in Illinois between 2002 and 2006 drawn from medical-record diagnosis codes. They showed how right Thomas had been to raise the alarm. Over those five years, hospital MRSA infections had risen 56 percent, from 6,841 cases to 10,714.[43]

By mid-2009, six other states had passed laws mandating counting MRSA cases, and twenty-nine had passed laws requiring hospitals to count and report infections in some manner. Seven states had already issued reports based on their hospital-monitoring efforts.[44] Illinois, though, was not among them. Its data was still held hostage, this time by a lack of funding for a new state computer system. The delay left Thomas, already engaged in lobbying for a new law to track MRSA in prisons and mental hospitals, furious and sad that preventing hospital infections could not remain a public priority for long.

"They have no idea what this does to families," she said. "A child gets it. The mother quits her job to care for him. They top out their insurance. They go into debt. And people are left for life with a disease they cannot shake. They become a shadow of what they once were."

Researchers still disagreed whether active surveillance prevented infections or justified its sometimes substantial cost, but they could agree on one thing: because surveillance identified people with previously unrecognized colonization, it increased the number of people who went into isolation. The good of the many came at a cost to the individual patient. Because of the time required to gown and glove, and the interruption in an already busy day that those tasks imposed, health care workers visited patients in isolation less frequently, and senior physicians less often than that.[45,46] Patients in isolation were less likely to get even basic care such as having their vital signs checked regularly by a nurse, and more likely to be left alone long enough for some adverse event to occur.[47] Many became lonely and clinically depressed, especially patients who were elderly or already psychologically fragile, who both needed more attention and were least likely to get it.[48,49]

The sacrifice could protect the health of others in an institution. But at the same time, as Jane McGuinn's family found, it could forever change the life of the person in isolation, and the patient's family as well.

The entry to McGuinn's family farm, known as the Mackin property since its 1865 founding, lies between two carved stone pillars and down a long lane lined with 150-year-old elms. The lane dips toward a rail-road track, climbs past a barn that dates back more than a century, and emerges on a hill overlooking 220 acres of corn fields, paddocks, and a forest of oaks threaded with a meandering creek. On the property's eastern edge, the grey of suburban asphalt fringes the horizon. On the western edge is the half-square-mile town of Leland, Illinois, population 970 and dropping. Inside the fences, the farm is otherwordly peaceful. There are horses in the sheds, turkeys in the cornstalk stubble, and barn kittens that play tag around a pair of somnolent corgis. The residents are Jane McGuinn, ninety-eight years old and the fourth generation to hold the land, her youngest daughter Molly Donahue, sixty-eight, and Donahue's husband John Pajkos.[50]

McGuinn was raised in Chicago, part of an affluent old manufac-turing family—her grandfather drowned on the luxury liner *Lusitania* when it was torpedoed in World War I—and she raised her children in the city as well. The farm, established by her great-grandparents on land they bought from the Chicago, Burlington and Quincy Railroad, was their country place, where cousins learned to ride and McGuinn trained jumpers and quarter horses that she and Molly showed across the Mid-west. Those were summer and weekend tasks. In her city life, she worked at a private school until she was eighty-nine, and then retired to the farm and ran a little free library with donated books.

The farm is beautiful, but remote, and as well-worn as any property used hard by many generations. The main house is a cottage that was added to over the years, with narrow stairs tacked to the outside wall and no insulation to hold out the steady prairie winds. By modern standards, it is not ideal shelter for anyone in her nineties, let alone someone with a glass eye and a replacement hip, who needs a rolling walker to steady her gait on the smooth plank floors. But the farm has turned into McGuinn's full-time home, and probably her last one, because isolation for MRSA during an unexpected trip to the hospital frightened her so badly that she is unlikely to ever leave the farm again.

Her MRSA troubles began in 2006. Four years earlier, McGuinn had had her hip replaced, and the joint still bothered her. Coming from a family with great teeth, low blood pressure, and a stoic indifference to minor injuries—ancestors who didn't die in disasters mostly made it to one hundred—she took an aspirin whenever she felt pain, and kept going. But then her side started to twinge. The aspirin had eaten a hole in her stomach. At ninety-six, she was carted off to a hospital for repair of a perforated ulcer.

The surgery went well, as did the recovery. McGuinn had no fever and no problems with the incision, and was chatty when her family came to visit and hungry when meals arrived. On her sixth day in the hospital, however, her daughter arrived for a visit and found an "isolation" sign taped to her mother's door. For the first time in her stay, McGuinn had been checked for MRSA—Donahue never found out what triggered the test—and had been found to be colonized. For her remaining four days in the hospital, everyone who entered her room would adhere to the hospital's MRSA precautions: wash, gown, glove, and mask; reverse the process on leaving, and wash again.

McGuinn needed additional recovery time—routine for someone of her age—and Donahue arranged a stay in a nearby nursing home. She lives in a Chicago apartment and teaches at the same school from which her mother retired, coming out to the farm on the weekends and school holidays. For a while her husband accompanied her, until one winter night when McGuinn left a door open and all the pipes froze. Donahue's husband moved out to Leland to help his mother-in-law, but the couple both hoped the arrangement was temporary. His wife was counting on the nursing home stay to provide a taste of assisted living that might bring her mother to consider moving away from the farm.

But the information that McGuinn was carrying MRSA followed her to the nursing home and dramatically changed the experience her daughter hoped for. The facility's infection-control regulations called for a single room, which was at the end of a corridor. McGuinn was not allowed to walk to the nursing station, or to the dining room. Food was brought to her on paper plates that were thrown away afterward. Anyone who entered the room was required to put on a gown and gloves, a significant barrier to any residents who might want to socialize, and an obstacle for staff with many tasks to do.

"They could never get there fast enough when she rang for help," Donahue said. "I don't think they were evil or careless, but they couldn't just pop in, so they would put her last because it was a chore to go to take care of her. So I would get there, and sometimes she would be in a wet bed, or there would be feces still in her commode. And her food was always cold."

In the two months she was there, to McGuinn's memory, no one but staff came into the room.

"I counted the trucks on the road, and trains, and birds outside the window," she recalled. "But they didn't keep the bird feeder full."

The frustration for the family was that McGuinn was never sick, and to their knowledge caused no sickness in anyone else either. Her cultures were inconclusive. She would show clear, but then a second confirmatory test required to lift the isolation order would indicate she was colonized again. At least two rounds of antibiotics and decolonization did not break the pattern, and there was no way to determine whether she was afflicted with a persistent bug, or being recontaminated by someone else.

McGuinn was trapped in her room, with no exercise and no social interaction. Seeing no resolution, Donahue brought her mother back to the farm, the home she had hoped they would all escape.

"Home is a fine place," Donahue said on a golden winter afternoon, sitting at the farm's dining room table and looking down the hill at a chestnut gelding nibbling hay. "But she's almost as isolated here as she was there. And assisted living is no longer possible." She raised her voice slightly, "Mother, would you live somewhere else?"

At the end of the table, McGuinn set her jaw and shook her head hard enough to shift the pins in her silver chignon. "I had enough of that," she said. "I'll just stay here."

Donahue sighed, and looked back out at the empty landscape. "That's the true cost of this," she said. "She's frightened of what happened, and of whether it will happen again if she leaves. So she won't leave. We won't ever get past that. And because she can't leave, neither can we."

Behind the struggle to control MRSA in hospitals and nursing homes lay a difficult reality: the bug persisted because it was so resourceful at evading antibiotics' control. In the decades since penicillin first failed to put curbs on the bacterium, drug after drug had been developed to fight it, and sooner or later the bacterium had developed defenses against them all. Science was failing against MRSA: It could find few biological mechanisms to exploit against it. And with resistance developing so quickly, the marketplace was failing as well. Fewer companies could persuade shareholders that antibiotic research would be worth their investment. After seventy years of the antibiotic miracle, the pharmaceutical pipeline was running dry.

174

CHAPTER 12

THE END OF ANTIBIOTICS

An air of anticipation lingered in the sleek suburban offices of Cubist Pharmaceuticals, in the high-tech corridor that wraps the western suburbs of Boston. It was January 2002. Behind the fanciful spire that marks the office's entrance, a lacy aluminum helix elaborated from the company's grid-like corporate logo, the staff was waiting for data that could open up the future for their ten-year-old company, or end it.

The small biotech shop, founded during the 1990s genomics boom by a Massachusetts Institute of Technology biochemist, had focused at first on patenting a method of identifying new molecular targets for antibiotics. Within a few years, though, the company leadership realized that the only thing that would earn enough money to make the firm viable was not a process, but a product. They needed to develop a new drug. Using work done and abandoned by another company decades before, they identified a new antibiotic, and quickly put it into clinical trials. Now they were about to "break the code"—to open the report that told them how well the recipients of their drug had done, versus the recipients of an existing drug that they had been required to compare it against.

So much was hanging on the results. The future of Cubist and its staff, many of them refugees from major firms that one by one had shut down their antibiotic-development programs; the fortunes of the drug, daptomycin, which had a novel mechanism for killing MRSA; and the possibility that small companies, and not the pharmaceutical giants of a generation earlier, could bring a new antibiotic to market. To some extent, even the future of antibiotic research itself depended on the results. Few companies desired to invest in antimicrobial drugs any longer, and every attempt that failed made the remaining practitioners rethink their research portfolios.

175

The code was broken and the data analyzed. The news was not good. Their drug had not done badly, but it had not performed well enough. The clinical trial had tested its effectiveness in community-acquired pneumonia. Daptomycin had cured 79 percent of the patients who received it; the comparator drug had cured 87 percent. The difference between the two was slightly more than 10 percent—but 10 percent is the maximum difference that the Food and Drug Administration allows between a new drug and a comparator when it decides whether a new drug should go forward to be licensed. In FDA terms, daptomycin had failed.[1,2]

"The in vitro data looked good," said Richard Baltz, PhD, Cubist's executive director of research, who had joined the company after decades in Eli Lilly and Company's research labs. "The animal data looked good. That it would not work, when you actually got it into a lung, was shocking."

There were all kinds of ironies in the findings. The FDA had just changed its standard from 20 percent to 10 percent; a year earlier, with the same data, daptomycin would have passed the test. The drug it failed against, Rocephin, had been on the market since 1984, and therefore had never been required to meet the same trial standard. FDA rules require new drugs for any condition to be tested against the best drug available, and for community-acquired pneumonia, Rocephin was considered the best. But Rocephin, also called ceftriaxone, is a beta-lactam antibiotic, which means it works against pneumococcal pneumonia, but would not be effective if it were given to a MRSA pneumonia victim. Daptomycin, envisioned as a new weapon against MRSA, had been judged less successful than a drug that MRSA had been resistant to for years.

The ironies were difficult to appreciate when the initial news was so unnerving. The rattled staff decided to press ahead. Daptomycin had already succeeded in one trial that had evaluated its use in severe skin and soft tissue infections, and another trial, for bacteremia and endocarditis, was scheduled for the summer. Perhaps the drug would succeed in the end, or perhaps run up against other unforeseen difficulties. Either way, Cubist's effort would demonstrate the scientific and regulatory challenges that new antibiotics can face.[3]

Daptomycin was discovered in the same place as most twentieth-century antibiotics—in a bit of something disgusting that came from somewhere far away. Starting with Robert Koch, who isolated the tuberculosis bacillus in 1882, microbiologists had recognized that common soil was simultaneously teeming with bacteria and lethal to disease organisms that were accidentally introduced to it. The promise of soil as a source of disease-killing substances also attracted the scientists Rene Dubos and Selman Waksman. In 1939 and 1944, they purified the antibiotics gramicidin and streptomycin from soil samples, in Waksman's case, from a farmyard where chickens had gotten sick.[4]

Starting in 1951, the pharmaceutical company Eli Lilly launched a massive effort to gather soil from as many far-flung places as possible. Lilly was looking for any organisms that made antibacterial compounds, but it was especially interested in ones that held the promise of killing staph. At the time, staph resistance to penicillin was building around the globe, and the bug had also developed resistance to new families of antibiotics such as erythromycin and tetracycline as they were introduced. As part of the program, headed by an organic chemist named Edmund Kornfeld, employees carried sterile sample tubes whenever they traveled away from the company's Indianapolis headquarters. Lilly also sent collection tubes and pleas for help across the planet, to the most far-flung friends, family, and distant connections that could be persuaded to dig up dirt on the company's behalf. The thousands of samples that came back were scrutinized for whatever promising compounds they contained. Sample 5,865, which arrived in 1952 containing mud from the jungle floor in Borneo, revealed the fungus *Streptomyces orientalis*, which ultimately yielded the basis of vancomycin.[5] Dirt scooped up in the mid-1960s on Turkey's Mount Ararat, the fabled resting place of Noah's Ark, contained *Streptomyces roseosporus*, the source for daptomycin.[6]

When vancomycin was isolated from the Borneo sample, it looked like a good anti-staphylococcal candidate. It was bactericidal, meaning that it killed organisms outright rather than just slowing down the growth of a bacterial population as some other antibiotics did. It was a new class of drug, a glycopeptide, which meant that bacteria had no prior experience in how to defeat it. Just as penicillin did, the new drug killed by interfering with the formation of a new cell membrane, but its mode of action was different: it bound to a different component of the developing cell membrane, in such a manner that bacteria ought to have a more difficult time foiling its activity. In "passage experiments," the technical term for growing generations of bacteria in the lab, Lilly researchers confirmed that resistance to vancomycin would be slow to develop. In twenty bacterial generations, vancomycin resistance increased only eight-fold. In an identically constructed test, penicillin resistance increased 100,000-fold.[7]

But vancomycin turned out to have challenges. It was difficult to purify, taking eleven separate steps to yield a liquid so sticky and dark that physicians called it "Mississippi mud." It was not absorbed into the system when given by mouth, and caused so much pain on intramuscular injection that developers concluded it would have to become an intravenous drug. Even by IV, though, it was less than perfect. It caused hearing loss, impaired kidney function, and painful inflammation in leg veins. It also did a poor job of pervading lung tissue and penetrating the central nervous system, making it unreliable for treating pneumonia and meningitis. Still, it passed muster in clinical trials, or what passed for trials in the days when the FDA's chief charge was determining whether pharmaceuticals were safe, not whether they worked.[8] Eight out of nine patients with a variety of staph infections recovered when given vancomycin, and so did five out of six patients who had endocarditis caused by a variety of organisms but had not been cured with other antibiotics.[9] That was enough for the FDA to fast-track approval, and vancomycin was licensed in 1958.

After its approval, however, it was almost immediately shelved. Its launch on the market coincided with the arrival of methicillin from England and of the other beta-lactams, all of which killed staph faster and more reliably, without vancomycin's side effects. Between 1958 and 1978, only a few hundred people around the United States received it, usually because they were allergic to penicillin-based drugs.[10]

But after 1980, MRSA's explosive growth in hospitals and then in communities created new demand for vancomycin. Between then and 2000, worldwide use of the drug increased more than a hundred-fold.[11] So much more use—usually in very sick people whose infections were resistant to other drugs—began to demonstrate something that physicians had suspected, but hoped was not true: vancomycin had more limitations than they had known. It took much longer to kill bacteria in the blood than the beta-lactams did, and it did not always quell some very serious infections for which it was the last desperate attempt at a cure. Even with vancomycin, some patients died.[12]

Vancomycin had achieved its status as the go-to drug for MRSA not because it was the best, but because it was the sole anti-staphylococcal antibiotic to which multidrug-resistant hospital strains were still vulnerable. From the time it was licensed, no vancomycin resistance had been recorded. But that was due in large part to the decades the drug had spent on the shelf unused. With so much more vancomycin suddenly in use, the resistance profile was bound to change.

The first change emerged not in staph, but in enterococci, bacteria that live in the gut and pose serious problems of wound contamination in hospitals. ICU patients in particular get many varieties of antibiotics, increasing the chances that their personal population of enterococci may become resistant to multiple drugs. The bugs get loose because antibiotics disrupt the ecology of the digestive system, and often cause diarrhea in debilitated patients. And if enterococci escape into the air and land on a surface that is not rigorously cleaned, they can survive there for days. By the 1980s, enterococci were already resistant to several classes of drugs.[13]

In 1989, the first case of infection with vancomycin-resistant enterococci, VRE, was discovered by chance in an elderly ICU patient in Baltimore. That the organism had picked up genes that conferred resistance should have been predictable and was probably inevitable. But many researchers had considered it unlikely, because disabling vancomycin's attachment to the cell membrane would be a complex process of several steps requiring a bacterium to produce several enzymes in a specific sequence. Enterococci had managed it, though, and in the process had rendered themselves virtually untreatable. Within four years, VRE invaded ICUs in thirty-three states, and accounted for 8 percent of all the hospital infections in the country.[14]

It seemed it would only be a matter of time before MRSA, already the most common cause of hospital infections, picked up vancomycin resistance and transformed itself from a challenging bug to an apocalyptic one. Critical care physicians felt an almost palpable tension, wondering which of them would be unlucky enough to find the first isolate of vancomycin-resistant staph in his or her ICU. But when vancomycin resistance did surface in staph, it was in an unlikely place and an unexpected manner. In 1996, Dr. Keiichi Hiramatsu—the leader of the Japanese research group that would later identify the *mec* cassettes used to distinguish community and hospital MRSA strains—found two patients at Juntendo University Hospital in Tokyo whose MRSA infections were not responding to vancomycin. One was a sixty-four-year-old man who developed pneumonia while recovering from surgery for lung cancer; the other was a four-month-old boy who had been operated on for a malformed heart valve. Both survived, but only after prolonged, difficult courses of treatment with several additional drugs, twenty-two days for the man and seventy-six days for the child.[15,16]

Vancomycin treatment in the two patients failed until they were both given unusually large doses. Vancomycin dosing had turned out to be something that hospitals did idiosyncratically, using formulas that computed how much to give by calculating factors such as weight, age, and the robustness of the patient's kidneys. As a measure that everyone could recognize, clinicians and researchers used the minimum amount of drug needed to kill bacteria, which they called the minimum inhibitory concentration, or MIC for short. The man's MIC, 3 mg/L, was at the high end of normal, which ranged from 0.25 to 4 mg/L, and the baby's 8 mg/L was far beyond it. Doctors believed that the threshold for completely untreatable was 32 mg/L.

With apprehension, Hiramatsu and his colleagues assumed that they knew what had happened: MRSA had imported the genes that conferred vancomycin resistance, probably from VRE. But when they ran an analysis, they discovered that assumption was wrong. Neither patient's MRSA strain harbored the vancomycin resistance genes that had developed in VRE. Instead, something else was happening. This MRSA with reduced sensitivity to vancomycin looked different under a microscope than MRSA usually did. It had somehow changed its structure; the walls of its cells were different, bulky and thick. For a drug whose mechanism depended on disrupting a cell's membrane, that had to be important— but the researchers were not sure how.

The community of staph researchers dubbed the odd new Japanese strains "vancomycin intermediate S. aureus," or VISA, and wondered whether they'd dodged a bullet, or whether more severe vancomycin resistance controlled by resistance genes would still emerge. They did not have long to wait. In June 2002, CDC investigators reported that a forty-year-old Michigan woman, a diabetic who was on dialysis for kidney failure, had a true VRSA infection, containing one of the genes that conferred vancomycin resistance in VRE. Her infection was in an ulcer on her foot that would not heal, a common problem in diabetics because of their poor circulation, Her MIC was an unheard-of 128.[17] Luckily, the bacterium was still susceptible to a few other drugs, and by December, the woman recovered.[18]

Within just a few months, there was a second VRSA case. An obese seventy-year-old man in Pennsylvania—also chronically ill, also with a foot ulcer—died eleven weeks after being diagnosed with the resistant strain. Researchers wondered whether they were witnessing the start of the epidemic they had feared—but microbiological analysis revealed that the two VRSA strains were not related. They both contained the key resistance gene, called *vanA*, but the internal evidence showed each had acquired it independently. Most important, in neither case did the super-resistant strain spread to anyone else.[19]

By the end of 2007, there had been nine VRSA cases in the world, all of them in the United States, and seven of them in Michigan. All of the isolates possessed the gene *vanA*. Their MICs ranged from 32, once thought to be the ceiling for untreatable, to an absurd 1,024. All of the patients were older, ranging from forty to seventy-eight, and all of them were chronically ill. But it was clear none of them had given the strain to each other. Genetically, it was evident that each strain had arisen on its own, as importations into the hospital strains USA100 and USA800 of three different genetic elements that had originated in VRE. For whatever reason—lack of evolutionary fitness, extraordinary infection control, or blind biological luck—none of the patients had sparked an outbreak.[20,21]

A VRSA crisis had been avoided for the time being, but it seemed that a VISA crisis was still building, with increasing numbers of infections caused by MRSAs that possessed mysteriously thickened cell walls. Though the mechanism was still not fully understood, it seemed clear that vancomycin locked onto the thicker membranes as it was supposed to, but that was all: at that point, the drug stopped working. Patients could still be treated successfully, but only if they were given higher doses of vancomycin that overwhelmed VISA's defenses. Then a third vancomycin resistance problem appeared, a much more subtle one: patients whose strains appeared normal but whose infections did not respond to normal doses of the drug. Painstaking analysis of individual isolates revealed why: within a patient's own population of MRSA, there would be clusters of bacteria that were individually more resistant, with much higher MICs than the surrounding staph. Researchers dubbed those h-VISA, for heterogeneous.[22] The aberrant strains seemed to explain a mystery that multiple medical centers around the country had been reporting, a phenomenon of patients with ordinary-looking MRSA infections requiring high doses of vancomycin. Doctors were calling it "MIC creep."[23]

The overall import was clear. Relentlessly, with the blind persistence of evolution, staph was trying multiple routes for escaping vancomycin's control. If the bacterium succeeded before other adequate drugs were brought to market, the result would be strains that were essentially untreatable. If it passed its new defenses to USA300—already virulent, multidrug-resistant, communicable, and ubiquitous—the result could be a superbug that was truly uncontrollable.

When vancomycin began to reveal its vulnerabilities in the 1980s, scientists at Lilly pulled out daptomycin, shelved for more than a decade, and ran tests on it. They found the compound had a shape that no one had recorded before: it was a molecule that looked like a lollipop, a ring of thirteen amino acids trailing a chain of lipids and carbon.[24] And that was exciting, because if the molecule had never been seen before, then it was likely to act in a manner that pharmacology—and more important, staph—had not yet experienced.

They were right. The molecule, a lipopeptide, had a unique mechanism. Up to that point, antibiotics had worked in two main ways. They either destroyed a bacterium by disrupting its outer membrane—that was how the original penicillins, the beta-lactams, vancomycin, and the cephalosporins worked—or they penetrated inside a bacterium, interrupting synthesis of DNA or RNA or proteins, which was the mechanism of macrolides, tetracyclines, fluoroquinolones, and rifampin. Daptomycin, though, bound to the outside of the membrane, changing the chemical gradients across it and depolarizing it. The bacterium could not generate energy or maintain its internal environment. Essentially, it lost its power supply and died. In an important difference, bacteria killed this way did not lyse, or break apart, as they did under the influence of membrane-disrupting drugs. So there were no multiple fragments whose presence could overstimulate the immune system into the kind of reaction that triggers septic shock.[25]

Recognizing daptomycin's mechanism of action allowed the Lilly researchers to determine which class of bacteria it could be used against: only the Gram-positives (so-called because they react in a particular way to a stain invented in the nineteenth century by a scientist named Gram), which include staph and enterococci. The Gram-negatives, a class that includes salmonella, shigella, and E. coli, have a double-layer cell membrane that would not be vulnerable. The new compound was promising enough for Lilly to put it into trials for severe skin and soft-tissue infections and bacteremia. While the first round of results was acceptable, the company wanted to see whether the drug would work better at higher doses. They brought the exploration back to a Phase I trial, a trial that tests only a drug's safety in a small number of subjects, increased the dose, and gave it twice a day to five volunteers. Two reported muscle soreness and weakness, and a check of their blood work showed elevated levels of an enzyme that indicates muscle breakdown. It was enough to change Lilly's mind. In 1991, daptomycin went back on the shelf.

"Most of the scientists and physicians working on the drug thought it still had promise," said Richard Baltz, who was one of them. "There was always a feeling that an opportunity had been missed."

Fast-forward five years, and Baltz was walking the very long hallways of the New Orleans Convention Center, site of the 1996 annual meeting of the American Society for Microbiology. Among the many posters on new pharmaceutical research, Baltz spotted one that looked interesting. A crowd of people had queued near it, waiting to talk to the author but not talking to each other. Such silence was a rule at scientific meetings, unwritten but nevertheless rigorously observed. You watched what you said, because you never knew when the person standing next to you might be a competitor.

Baltz waited his turn, arrived at the front of the line, chatted for a few minutes and shook hands. As he turned to go, the poster's author inexplicably broke protocol and drew another bystander into the conversation. "I don't know if you know each other, but this is Frank Tally," he said.

The two men had never met, but it was impossible for Baltz not to know Dr. Tally's name. Among drug developers, Tally, who turned fifty-six that year, was legendary. Pharmaceutical biochemists count themselves lucky if they can get one drug through the pipeline, from development to trial to market, in their professional lives. In nine years at Lederle Laboratories after a first career as a physician, Tally had already supervised three: cefixime, tigecycline, and the combination piperacillin/tazobactam, at the time the bestselling injectable antibiotic in the world.[26] It was widely known in the pharma world that he had left Lederle the year before to join Cubist Pharmaceuticals, and it was expected that he was looking for a fourth good drug.

By the time their chat ended, Baltz had agreed to give a seminar at Cubist, an appearance that was tacitly understood to be an audition for a future job. By the time Baltz finished the visit months later, Tally had fixed on daptomycin as Cubist's first drug candidate. In 1997, Cubist licensed the compound from Lilly, and started research.

Tally's move was not what the pharma gossip networks had expected. "The early line was that this was an irretrievably toxic drug," said Jared Silverman, PhD, who had just joined Cubist as director of microbiology. "And that we were obviously idiots."

* * *

Ten or twenty years earlier, when Baltz was working for Lilly and Tally was working for Lederle, it would have been unthinkable for a small company such as Cubist to develop an antibiotic. Since then, however, the economics of pharmaceutical manufacturing had changed. Lilly and Lederle—along with Aventis, Bristol-Myers Squibb, Roche, and Procter & Gamble—had shut down or sharply curtailed their antibiotic research efforts.[27,28] (Lederle was later folded into Wyeth Pharmaceuticals.) Pharma companies were deciding that antibiotics were no longer guaranteed moneymakers. The drugs are expensive to develop; the full cost of researching, licensing, and marketing one is estimated to be $800 million.[29] Once on the market, antibiotics face competition not just from already approved antibiotics, but from the chronic-care drugs for diabetes, kidney failure, and heart disease that eat up an increasing percentage of health care spending.[30]

Add in that, if a bacterium is susceptible to an antibiotic, a patient will take the drug only for several weeks; or for several months, if the organism is resistant and the infection is prolonged. That is a limited amount of consumption during which to earn back an investment, compared to the amount of Prozac or Wellbutrin that a depressed patient might take daily for years, or the Viagra or Cialis that a healthy man might take for decades.

Finally, factor in that organisms begin to develop resistance as soon as an antibiotic is put into play against them. The effective marketplace life of a new antibiotic is unpredictable. It took forty years for staph to develop resistance to vancomycin, but the first clinical proof of MRSA's resistance to linezolid, another new compound with a novel mechanism, came within a year of that drug's FDA approval.[31] And prescriptions decline once tests begin to show that 15 percent or more of an organism, in a hospital or a geographic area, is resistant to a new drug. Physicians become reluctant to trust it, fearing they run too great a risk of patients not being cured.[32]

"It is the prevailing view in Big Pharma that there are better ways to invest research dollars," a Wyeth scientist bluntly warned in a medical journal in 2003. "Quite simply, if we want new therapies then someone is going to have to pay for it."[33]

Antibiotics move through the federal approval process faster than other classes of drugs, and are more likely to be approved at the far end, two factors that ought to balance out the cost of drug development.[34]

184

Nevertheless, between 1980 and 2003, though the fifteen largest drug companies brought 418 new drugs into the regulatory process, only five of them were antibiotics. None of the five had a new mechanism; they were refinements of existing molecules. The companies were not backing away from all research, though. Their total research and development budgets went up over that period, not down, and they brought forward sixty-seven new drugs for cancer, thirty-three for pain, thirty-four for metabolic or endocrine disorders, and thirty-two for pulmonary disease. It was only antibiotic research they were abandoning.[35]

In December 2008, the Infectious Diseases Society of America identified eight potential anti-MRSA antibiotics under development in the United States, though only three of them were in the final round of trials that would bring them close to market. Within a month, two of the eight were withdrawn from the licensing process, and a third was put on hold for another round of trials.[36]

At the same time, physicians taking care of MRSA patients were growing impatient. They needed new drugs immediately. Often patients were sick with several things—chronic illnesses, diseases of old age, and infections incurred in the hospital—and the more drugs they were taking, the more likely it became that they would suffer a cross-reaction and need to be switched to something else. The fewer drugs there were available, the more difficult such swaps became.[37] Furthermore, some of the new drugs were exhibiting unexpected toxicities that hadn't been spotted in trials. In one hospital in Australia, 64 percent of patients who were on long courses of linezolid because of serious infections experienced side effects that its trials had not uncovered, from abnormal blood counts to nerve damage in their legs and feet.[38]

"It is extremely rare, when we have a discussion about how to treat a patient, that we do not have to run through a list of one or two or many antibiotics that have not worked for them," said Dr. John Edwards Jr., the chief of infectious diseases at Harbor-UCLA Medical Center in Los Angeles. "When we are on rounds, we spend the vast majority of our time going from one patient to the next who are infected with organisms for which we have almost nothing that works, and trying to figure out how to treat them."[39]

At Cubist, the thought at first was that daptomycin could be a topical antibiotic against MRSA skin infections, or an oral drug that would send its potent antibacterial action into the gut to combat resistant enterococci. Frank Tally, though, thought the drug had been misunderstood. He and Cubist toxicologist Rick Oleson launched a set of experiments in dogs to get at the roots of daptomycin's muscle toxicity. What they discovered was counterintuitive and demonstrated how tricky deploying a new compound could be.

Most antibiotics, including the beta-lactams, rely on keeping an even level of drug in the blood over an extended period of time. To reach that level and stay there, doses are given two, three, or even four times a day. That killing effect is called time-dependent. But daptomycin, by contrast, was concentration-dependent: for maximum effect, an entire day's worth needed to be given at once. The drug was slightly destructive to muscle, as the original testers had suspected, but with a once-a-day dose, that was unimportant. Daptomycin had an eight-hour half-life; at eight hours, half of the drug had left the system, and what remained constituted a low enough level in the blood that the body could undertake muscle repair before the next dose arrived. Where Lilly had gotten into trouble was in keeping the dose even, so that muscle damage built cumulatively and without any drug-free time in which the body could make repairs.

That realization made it safe to use daptomycin as an IV drug as Baltz and the other developers had first envisioned, a delivery method that would allow them to apply it against the most serious infections. Cubist secured a patent on the dosing method—it was the only novel intellectual property available, since the patent for the compound itself had expired years earlier—and set up several trials, in skin and soft tissue infections, community-acquired pneumonia, and endocarditis.

When the news came that the pneumonia trial had failed, the researchers could not make sense of the result, and it was vital that they understand it quickly. The drug's behavior in this trial could determine whether it succeeded with endocarditis, one of the next trials scheduled, because an infected heart valve can shower septic emboli into the lungs. If daptomycin did not work in the lungs, it would end up as just a drug for skin infections—not scoring the big win the company wanted, and not contributing to medicine's most important needs.

Results of pneumonia studies in animals, done in the 1980s, had been sent with the compound from Lilly. Director of microbiology Jared Silverman decided to repeat them, using mice and rats. When he got results from his lab, he couldn't make sense of them either: in the lab animals, the drug did not affect pneumonia at all. The surprise was not that the drug had not succeeded in Cubist's human trials, but that it had done as well as it did.

After multiple rounds of animal tests, using different organisms to mimic different types of pneumonia, the lab team finally figured out what had gone wrong. The problem was pulmonary surfactant, a waxy mixture of fats and proteins that coats the air-exchange surfaces in the lungs to regulate surface tension and to keep them from sticking together on exhale. Surfactant, it turns out, is remarkably similar in composition to bacterial membrane. When daptomycin entered the lung, it mistook the surfactant-coated surfaces for bacteria and bound to them, using up the drug before it could attack the organisms it was aimed at. The Lilly experiments had completely missed that vulnerability, because to make sure the infection in the research animals presented a robust target, the company's researchers had infused an enzyme into the animals' lungs that traumatized the tissue. The enzyme had broken through the surfactant, allowing daptomycin to work, but only in an artificial setting.

The finding contained both bad and good news. It indicated that daptomycin could never be a treatment for MRSA and other bacterial pneumonias, because there would be no way to bypass surfactant's inadvertent lure. But it also cleared the drug as a possible endocarditis treatment, because septic emboli would be within the bloodstream, not on the surface of the lung.

In September 2003, the FDA approved daptomycin for treating complicated skin and soft-tissue infections, based on results in 1,400 human patients and lab studies on 21,000 bacterial samples. Like the pneumonia study, this one was against comparator drugs, because it would be unethical to knowingly give a placebo to someone with a serious infection. Daptomycin fulfilled the FDA mandate of being "not inferior" to its two comparators, vancomycin and the beta-lactam nafcillin. It achieved significantly more cures in streptococcal infections, and was within a few percentage points of the comparator cure rate in MRSA.[40,41]

Meanwhile, the enormous endocarditis trial, suspended for seven months while pneumonia data was analyzed, screened more than 5,000 patients at forty-four medical centers in four countries to find the 235 patients that were eventually enrolled.[42] The last patient finished treatment in February 2005, and in March 2006, the supplemental application to approve daptomycin for treatment of bacteremia and endocarditis came up for a vote before the FDA's Anti-Infective Drugs Advisory Committee. A decision was clearly needed: according to company data,

150,000 patients had received daptomycin in the two and a half years since the first approval, and probably one-fourth of them had been off-label treatment for bacteremia that was not treatable by existing drugs.

The patients in the trial were very sick, not only with bacteremia, but with diabetes, heart disease, and recent surgeries—21 percent were injection-drug addicts and 4 percent were HIV-positive. The rates of cure, determined by an independent committee, were 44.2 percent for daptomycin, and 41.7 percent for the comparator drugs. The number of enrolled patients who could be proven to have endocarditis turned out to be small. The FDA committee members pressed the Cubist representatives with questions, about stubborn bloodstream infection, the potential for daptomycin MICs to rise, and deaths among participants. In the end, however, they recommended the FDA approve the drug, voting 9–0 to license daptomycin for treatment of bacteremia and splitting 5–4 for endocarditis on the right side of the heart.[43]

The license was officially granted to Cubist in May 2006, and in September, the team prepared for a major scientific meeting, ready to bask in congratulations for getting daptomycin passed. They especially expected attention for Frank Tally, who had achieved his fourth drug—but about a week before the meeting, Tally began complaining of a sore shoulder, something that had bothered him periodically since he dislocated it in varsity football decades before. Tally was sixty-six now, and not in perfect health. He felt unwell enough to be hospitalized, and told his fretting colleagues to go to San Francisco without him. While they were at the meeting, Tally died of bacterial endocarditis. It was a painful reminder that the organisms he had spent his life combating were deadly opponents still.

Daptomycin, his last drug, was his legacy: it has become another MRSA drug of last resort, used by IV in the most serious MRSA cases. But staph is relentless. By the summer of 2009, three years after the FDA approval for bacteremia and right-sided endocarditis, twenty cases had been reported in which daptomycin's MICs in bacteremic patients edged up from 0.25 to 4, and in one case to 8.[44]

Developing antibiotics was so challenging, and resistance so inevitable, that medicine began considering an alternative strategy: persuading people to use fewer antibiotics, so that resistance would be slower to arrive.[45] That ought to be easy, because large amounts of the antibiotics prescribed in the United States each year are unnecessary. But reducing antibiotic use, even pointless use, has proven remarkably difficult.

About 75 percent of the antibiotics prescribed outside of hospitals in the United States are for acute respiratory infections: colds, ear infections, sinusitis, bronchitis, and sore throats.[46] However, most of those infections are caused by viruses, and antibiotics do not affect viruses; they work only on bacteria. Physicians know this quite well, since no one would graduate from undergraduate biology, let alone medical school, without understanding the difference. Yet 44 percent of children's visits to the doctor for cold symptoms, and 75 percent of visits for bronchitis, end with an antibiotic prescription being written.[47] So do 83 percent of adults' visits for sinus infection and inflammation, a set of symptoms that are much more likely to be caused by viruses—or fungi or allergies—than by bacteria.[48] Overall, most of the antibiotics prescribed during the 76 million office visits made in the United States each year for respiratory infections have no effect on the organism that is making the complaining person sick.[49]

Numerous educational campaigns, aimed at physicians or patients or both, have tried to change pointless prescribing, with limited success. In the 1990s, the CDC created a nationwide campaign, "Get Smart: Know When Antibiotics Work," and in 2002, the agency estimated that prescriptions for children's respiratory infections had decreased 14 percent nationwide.[50] Separately, an aggressive year-long educational campaign in Tennessee forced unnecessary prescriptions down by 11 percent between 1997 and 1998, and a Colorado campaign in 2001 cut adult prescriptions for viral bronchitis by half—but the percentage of visits in Colorado that ended with a prescription being written could not be reduced below 36 percent. Because they were for a viral illness, microbiology dictated the percentage should have been zero.[51,52]

Educational campaigns had some impact, but behavior did not change for long. The percent of bronchitis patients to whom physicians gave antibiotics fell from 77 percent in 1995 to 59 percent in 2000, but it had bounced up to 67 percent by 2005.[53] Ominously, most of the increase was in the new drug classes that infectious disease experts wanted to protect for as long as possible to preserve their potency. Between 1990 and 2000, prescriptions of the relatively new macrolides azithromycin and clarithromycin went up 368 percent, and between 1995 and 2005, prescriptions for the newest fluoroquinolones went up 128 percent.[54]

When researchers ask physicians why they write so many unnecessary prescriptions, doctors reply that patients want them. In one of the most-cited studies, the Alliance for the Prudent Use of Antibiotics asked 499 primary-care doctors in Massachusetts why they prescribed antibiotics when they did, and 59 percent replied that it was because patients expected them—twice as many as said it was because they were under pressure to get patients out the door quickly, and three times as many as said they would be sued if they did not.[55] But doctors' perceptions of their patients' desires may be mistaken. In 2001 and 2002, two Harvard physicians asked 310 people who came to an urgent care clinic with bronchitis whether they had intended to ask for antibiotics, and found that while 46 percent of those who did want prescriptions got them, 29 percent of those who had not intended to leave with a prescription did.[56]

The difficult reality is that giving a patient a prescription signals that something has been accomplished during a doctor's appointment. Medicine has yet to find a substitute action that sends the same signal. To curb inappropriate prescribing, pediatricians' national practice guidelines now call for not prescribing antibiotics immediately if an ear infection is mild and not accompanied by fever, and waiting for a few days to see whether the problem resolves on its own.[57] Even though most ear infections heal in that time—only a small percentage are actually helped by antibiotics—in a survey of several thousand Massachusetts parents, only two-fifths said that having a doctor prescribe "watchful waiting" would be an acceptable way to handle their child's care.[58]

One place where antibiotic use might be effectively curbed is hospitals, because the institutions can exert control over their patients. But such a practice would require striking a balance among needs that can be in conflict: choosing the right drug for an infection, not contributing to whatever resistance is emerging in the institution, and avoiding drugs that are expensive if cheaper ones will do the job.

There are health care industry recommendations, collectively called "antibiotic stewardship," for what hospitals should do to reduce inappropriate prescribing and recognize and combat emerging resistance. Published in 1997 by the professional organizations the Society for Healthcare Epidemiology of America and the Infectious Disease Society of America, they include monitoring drug resistance and drug use in the hospital, maintaining a list of approved drugs for the hospital pharmacy tuned to the hospital's own resistance patterns, using clinical guidelines that recommend specific drugs for specific diseases, and using mechanisms to divert drugs from use when necessary, such as restricting antibiotics from being dispensed or including automatic "stop orders" that force physicians to evaluate whether a patient can be switched to a more narrow-spectrum, less resistance-inducing drug.[59]

Within just a couple of years, evaluations by the two professional societies showed that while most hospitals were using at least one of those strategies, none were using all of them. Doctors complained that they did not have enough information available to them at the moment they were making prescription decisions, and physicians battled among themselves and with hospital administrators over how much autonomy they could have in treating their patients.[60,61] Meanwhile, antibiotic use in hospitals kept rising, with antibiotics that work against a number of organisms, and especially vancomycin—the ones medicine most wants to protect from resistance—becoming the most-used drugs.[62]

"The individual physician and the patient, they don't care about resistance, about cost, about breeding superbugs," said Dr. Christoph Lehmann, a neonatologist at Johns Hopkins University Children's Center. "They care about this one individual. The system is set up to be full of conflict."[63]

Lehmann, who also specializes in health informatics—the science of extracting and manipulating data from computerized medical records— was sitting in one of the tiny offices where academic medicine seems to stash its most innovative researchers, a ten-foot by twelve-foot room at Hopkins that gets almost as much light from computer screens as it does from its airshaft window. While working in newborn critical care, Lehmann has launched more than a dozen computer-programming projects, from automated calculators and order-entry programs that reduced medication errors, to an online atlas of dermatology images and a Web-based teaching program that leads physicians through a first encounter with a virtual patient.[64]

One of his projects came about when his friend Dr. Allison Agwu, a pediatric infectious diseases specialist, was complaining how often she was paged to authorize a prescription on the hospital's restricted list. In the 1980s, as an attempt at stewardship, Hopkins had created a "prior approval" system that required physicians in other specialties to page the pediatric infectious disease fellows. The fellows would call the physicians back, explore the case, rule on the drug request, and notify the hospital pharmacy that dispensing the drug was okay. More than 1,400 such calls were made each year. The fellows tried to use the calls as education, leading the other doctors through a discussion of what drug would work best for the situation, but they often found themselves telling the same doctors the same thing.

"I was getting bombarded with calls from people that always ended with, 'Oh, you told me this earlier—I forgot,'" Agwu said. "So I started writing down all the approval requests, just to catalog them, along with the rationale for what I decided, and giving copies of the sheets to the pharmacy as well. After a while they started asking for them, and I thought, *There's a need here.*"

At the same time, Lehmann was thinking about converting the phone-based approval system—despised by almost everyone who used it, from fellows receiving too many pages to nurses who complained that drug doses were delayed and missed—into a computerized one.

"There was constant gaming," he said. "People would wait to put in orders until after eleven o'clock, because there was automatic approval overnight when the fellows were not on duty. Or they would look up which fellow was on call on which day and then plot to resubmit the order to the easygoing one. And nothing kept track, so you could have a fellow approve one week's worth of a drug, and three weeks later the patient would still be on it."

In the twenty years since the approval system was created, Hopkins had introduced electronic medical records that could be accessed anywhere in the hospital. By tying a new approval program into those records, Lehmann realized, he could have the system automatically generate a census of current patients, with their diagnosis and the drugs they'd already received, and include decision-support coaching that would start from the diagnosis and reproduce the conversations Agwu kept having to repeat. There would still be a human component—once the computerized request was completed, the system would page the on-call fellow to review it—but there would also be a history that documented any attempts to manipulate the system, as well as any prior prescriptions that the physician might have neglected to mention.

Ordering prescriptions through a central computer system rather than on paper was not a new thing in hospitals. For about a decade, it had been part of a nationwide push to reduce medical errors by removing haste, misunderstanding, and physicians' notoriously bad writing from the process of getting patients the right drugs and doses. But tying antibiotic stewardship to what hospitals call "computerized physician order entry" was relatively new. Additionally, some hospitals had attempted to control outbreaks of specific resistant organisms by dialing in stewardship directives that governed use of specific drugs. But Lehmann's broad approach, challenging every antibiotic order as it came in, was new and bold.

Agwu and Lehmann collaborated with pharmacy colleagues and in June 2005 launched a Web-based program that led the person requesting the drug through the complete approval process—decision coaching, page to the fellow, discussion if necessary, and pharmacy notification—before any order was placed. After a year, the new program had done a better job curbing use of the most precious drugs, measured by number of doses dispensed (down 11 percent) and amount paid for them (down 21 percent). Fewer doses were missed or delayed, and the number of phone calls to ask for restricted drugs sank 40 percent. Best of all, health care workers didn't mind using it. Physicians and pharmacists said they were three to five times happier with the new system than the old one.[65] Agwu, now an assistant professor at Hopkins, found she was having more conversations that discussed the subtleties of prescribing, and fewer that turned on the balance of power between the ordering physician and herself.

"It's not so confrontational anymore," she said. "It's . . . Hmm, I didn't know this was not the only drug to use."

Ironically, one thing they couldn't determine was how much the enhanced stewardship program suppressed the development of resistant organisms. They had faith that it had done so, but there was no way to provide proof.

"It's too complex," Agwu offered. "It's affected by what the infection control department is doing, whether the isolation policies are being enforced or have changed, what the environmental department is doing. Resistance is produced by a system, and we are only one component in the system."

"There are too many confounding factors," Lehmann agreed. "But we know for instance that after we implemented this, patients' days on restricted drugs went down, and patients' days on unrestricted drugs did too. That was something we did not anticipate. So we are seeing significantly less use of antibiotics overall, and we believe we are creating the conditions that will have a positive effect."

Lehmann and Agwu were not the only researchers to discover that antibiotic use could be reduced in creative ways that controlled costs and emotions and may have reduced resistance. Other institutions, from Australia to Utah, had made attempts at similar systems.[66,67] Most had found the systems worked as planned—but every case was a one-off whose success depended on the initiative of staff members who were willing to design the system, put it into practice, and campaign for it against skeptical or overworked colleagues. It was only incremental progress when broad change was required. The need to control resistant organisms' exposure to antibiotics, especially in the pressure cooker ecology of a hospital ward, was greater than ever. Because, as an epidemiologist in New York City and a mother in Atlanta were about to find out, MRSA infections in hospitals were taking a new and alarming turn.

CHAPTER 13

THE EPIDEMICS
CONVERGING

Diane Lore was about to give birth, and she thought she was prepared. She would be having her third child, and like the other two, the new baby would be born by prearranged Caesarean. Her nine-year-old son, the oldest, was packed for camp near his father's house in South Carolina. Her second husband Richard Ross had switched to night shifts at UPS so he would be around to assist during the day. His parents were ready to arrive in the Atlanta suburbs where the couple lives, eager to help distract their three-year-old. Lore, who was three months from her fortieth birthday, was the medical editor of the local newspaper, so she knew the reputation, the metrics, and most of the doctors at the hospital where her delivery was scheduled.

Nothing, she felt, could go wrong.[1]

Emily Rose Ross, her daughter, was born as planned on June 6, 2003. The delivery was uneventful, but the nurses who whisked the newborn off to check on her noticed that her blood sugar was low and asked Lore to start breast-feeding immediately. Emmy clamped down so hard that she gummed a tear in her mother's right nipple. Within a day, it started to hurt. When Lore was discharged four days later, her breast was so sore that breast-feeding was unpleasant. As she checked out, a lactation consultant told her to buy ice packs and slip them in her bra.

At home, Richard and his parents had cleared out the oldest child's upstairs bedroom. It was meant to be a temporary nest while the C-section healed and mother and baby got to know each other. The plan was for Lore to carry the baby into the household swirl when she felt up to it, but after four days at home, she couldn't get out of bed. Every time

Emmy locked on her breast and pulled, she felt a lancing agony that left her folded up and weeping. In between was not much better. She was feverish and weak, and her breast was more swollen than it should be and so tender she couldn't bear to touch it. She had breast-fed both of her other children and she'd never experienced anything like this.

She called every member of the local La Leche League chapter from bed, working her way down the list of fifteen names; disordered by the pain, she left frantic messages for all of them and then started at the top of the list again. Her lactation consultant diagnosed an infection and tried to pull it out with a breast pump. When Lore showed the breast to Emmy's pediatrician at the baby's two-week checkup, he thought it looked so alarming he urged her to switch to formula. She wangled her way into her primary care physician's office after hours, churning across town in Atlanta's gridlocked traffic. He diagnosed mastitis and possibly shingles, and he prescribed painkillers, an antiviral, and cephalexin, a beta-lactam cephalosporin. She had a post-C-section checkup with her obstetrician, who told her to see a surgeon immediately and prescribed more potent painkillers and azithromycin, a macrolide. The surgeon she found gave her amoxicillin-clavanulate, a combination that includes a penicillin derivative.

None of the drugs Lore took affected the infection. It continued to spread. On June 29, three weeks after Emmy chomped her and one day before her surgery appointment, she woke up with chest pains, unable to move her arm and shoulder on the left, uninjured side. The skin on the infected side was bright red and blistering and she felt as if her breast would burst. Richard took her to the emergency department. There was a three-hour wait that Lore spent in a trance of pain, while Richard, alarmed by the chest pain, fretted she was enduring a heart attack. Eventually they were conducted into a room by a physician's assistant. Richard helped her get her top off and expose the infected breast.

"Holy Christ," the PA said.

When the attending physician arrived he told her it needed to be lanced, but since her surgery was already scheduled for the next day, he was going to advise her to wait. After loading her up with potent narcotics, he sent her home. The next morning, the surgeon punched three holes into her breast. Enough pus leapt out to fill two irrigation trays. The doctor sent some for culture, and pushed a wick, a thick piece of cotton batting, through the holes in her breast to help the infection drain.

Three days later, the doctor called to say the infection was MRSA.

None of the antibiotics she'd taken in the past month had done her any good.

The physician sounded surprised, and he had good reason. Staph mastitis is not very uncommon—about one out of twenty women might be unlucky enough to get it—but in 2003 MRSA mastitis was almost unheard of.[2] Pregnant women have almost nothing in common with the elderly or debilitated, long-hospitalized, or surgical patients who are the usual victims of hospital-acquired MRSA. Evolution has arranged for them to be robustly healthy, and under modern managed care they are seldom in the hospital for long. Lore's case was doubly unusual; based on the timing, her infection had occurred in the hospital around Emmy's delivery—but based on the drug-sensitivity analysis, she'd been infected with a community strain.

As far as her doctors knew, cases like Lore's just didn't happen, One group of physicians in the United States, however, knew differently. Once again, far in advance of the public's knowledge and most of the medical community's awareness, MRSA's behavior was changing.

The September before Lore's ordeal, an obstetrician in New York City had uncovered something equally odd. A woman who'd just given birth was complaining of a serious breast infection. Her doctor prescribed beta-lactam antibiotics, but the mastitis had not responded.

The obstetrician called Dr. Lisa Saiman, a pediatric infectious diseases specialist at Columbia University Medical Center and the epidemiologist at New York–Presbyterian West Campus, where the woman had given birth. She said a phrase that always makes an epidemiologist's heart race: "This is weird."[3]

The woman's infection was not responding to treatment because, against all medical experience, it was not caused by drug-sensitive staph. It was caused by MRSA. Dr. Saiman understood how rare that finding was. For a postpartum woman to contract MRSA in a hospital, many other things would have to be aberrant first, from her underlying health to her treatment to the length of her stay. Other than her infection, though, this woman was completely healthy. Adding to the anomaly, lab analysis showed that her infection could not be due to the hospital strain of MRSA. It was resistant to beta-lactams and erythromycin, but susceptible to an array of other drugs.

There was an eerie historical resonance in the finding. New York–Presbyterian West Campus was the teaching hospital of Columbia

University Medical Center, the same hospital where a confused cab driver had taken Beverly Dieringer, addled with fever and staph mastitis, almost fifty years earlier. In the decades since, crosstown rivals Columbia and Cornell had started sharing Manhattan's medical institutions. Their joint educational efforts were based at the other New York–Presbyterian, on the east side of Manhattan—which once had been called the New York Hospital, and was the place where Heinz Eichenwald and Henry Shinefield solved the puzzle of S. aureus 80/81.

At about the same time as the obstetrician called her, Saiman heard from the hospital's clinical microbiology lab, which ran routine tests to detect the emergence of any resistant organisms. The lab had two other isolates from women who had given birth in August 2002 and returned with serious skin infections. Both isolates were MRSA. They too exhibited the resistance patterns of a community strain. After running PFGE on the isolates, the lab compared the result against a database, and discovered something unprecedented: all three women had MW2, the strain that had infected Robert Daum's earliest patients and killed children in Minnesota and North Dakota in the late 1990s, the strain that the CDC was beginning to call USA400.

Saiman and the lab were completely puzzled. In 2002, community MRSA had barely reached New York City.[4] It had certainly never been seen in a hospital there. But three cases in one hospital, in women who had nothing else in common, were not just an anomaly; they were an outbreak. As far as Saiman knew, an outbreak of community MRSA in postpartum women in a hospital had never happened before.

"As an epidemiologist, you are taught to be alert for abnormal patterns," Saiman said. "This was a pattern different from anything that we—that anyone—had seen before."

She and the rest of the hospital's epidemiology department were so startled by the discovery that they launched a full-on search of New York–Presbyterian. They sent letters to every member of the obstetrical faculty and every outside obstetrician who had privileges to work at the hospital. To raise awareness, they staged a Grand Rounds, a lecture-style gathering of physicians that explores specific cases and clinical trends. They asked the microbiology lab to double-check any staph isolates that had come from the ob/gyn service in the past year. In the end they found there were eight women: four with mastitis, one with an infected C-section incision, one with cellulitis, and two with pustular rashes. One had a MRSA urinary tract infection as well. Five of them had been

198

sick enough to go to other hospitals for treatment some time after they had gone home with their babies. All eight had delivered at New York Presbyterian–West Campus in the same two-week period in August, and they all had the community strain USA400.

Saiman and her colleagues couldn't make sense of the outbreak. They organized a mass swabbing and culturing of all 178 employees who had had contact with the women or their babies, even though none of the babies was sick. Not one employee was carrying MRSA. Pressing on, the team checked 100 newborns brought back to the hospital for healthy-baby checkups during September, to see whether babies were being colonized with MRSA there. None harbored MRSA, though four were carrying drug-sensitive staph. In a long shot, they swabbed environmental surfaces in the labor and delivery area, while admitting to themselves that a month of routine cleaning would probably have wiped clean any lingering contamination. They found no MRSA, but did find one strain of drug-sensitive staph.

When Saiman reconstructed the outbreak, she could see only one possible link, and it was so tenuous—suggestive, but unprovable—that she left most of the details out of the medical journal article in which she recorded the outbreak's course. Saiman identified two key time periods in the outbreak: the two weeks during which all the women delivered in the hospital and presumably picked up the bug, and a much wider period, more than two months, in which they began falling ill. She made a graph with a timeline, and on it plotted each of the major events—hospital admission, hospital discharge, onset of symptoms—for each of the eight women. When she did that, one element stood out: one of the eight women had been in the hospital twice during times in which the others had come and gone. She was the only common element to whom all the other seven new mothers had been exposed.

The woman was the seventh of the eight to give birth during the two-week window; call her patient 7. She had been hospitalized in obstetrics in the first week of August for preterm labor. While she was there, patients 1, 2, 3, 4, and 5 checked in, delivered their babies and were discharged. Patient 7, still pregnant, was discharged as well. Within a few days, though, she came back. She had placenta previa, a serious birth complication in which the placenta covers the opening to the birth canal and can bleed heavily as the uterus expands. Her baby was delivered by C-section, after which she had a hysterectomy. She was in the hospital for two weeks after the surgery, and in that time her C-section incision became infected. While she was there, patients 6 and 8 came, delivered, and went home. After she checked out, no more women acquired MRSA on the obstetrics ward.

The outbreak left Saiman with the clear sense that MRSA's parallel epidemics were converging—because here was a well-defined outbreak that spread within a hospital, but was caused by a community strain that had come from the outside world. She had the unmoored sensation that even the terms commonly used for MRSA—community-associated and hospital-acquired—would not be applicable much longer. She was uncertain: had New York–Presbyterian witnessed a brushfire, or the first spark in a conflagration that would sweep health care as staph 80/81 had done?

There was no wildfire yet but, other sparks had begun to land elsewhere in the city. In October and November 2002, six newborn babies and a woman who had just given birth were diagnosed with MRSA at Maimonides Medical Center in Brooklyn. They all had MW2.[5]

In Georgia, Diane Lore had received her diagnosis, but she was far from achieving a cure. The first hurdle was getting the appropriate drugs for the infection, which had a community-MRSA resistance profile—insensitive to penicillins, cephalosporins, and macrolides, vulnerable to everything else. Her surgeon referred her to an infectious-diseases consultant, who prescribed linezolid plus rifampin, mupirocin to decolonize her nose, and more painkillers. She received the prescription on July 3, a Thursday. Eager to get the drugs before the holiday, Richard hurried to a local pharmacy. They told him that linezolid, a new drug still under patent, would cost him $6,000 just to cover the weekend.

There was no way the couple could afford the drug, so they negotiated a change with the doctor, to trimethoprim-sulfamethoxazole, while their insurance company debated whether to approve the linezolid prescription. The sulfa drug gave Lore migraines, and it did not affect her symptoms. On Monday, July 7, she went back to the surgeon to have two more holes punched in her breast and another floppy, dangling drainage wick put in.

The MRSA diagnosis provided an explanation for her illness, but it threw their household into turmoil. Already, distraction and pain had kept Diane from bonding with her new daughter. She was too weak to hold Emmy and too drugged to sleep safely alongside her. Suddenly, she had to fear that she was an infection risk to the baby, her other children, her husband, her in-laws, and her parents—especially her mother, a partially disabled wheelchair user who had come to provide whatever support she could. Lore's maternity leave was about to end, and she needed to negotiate another six weeks of medical leave. She was tormented by how sick she was, and embarrassed by her symptoms.

"I don't think I was vain at any age, and I don't think at forty after three children my breasts were all that fantastic," she said. "But I was so horrified by how they looked, what had happened to them—when my mother-in-law and sister-in-law would change the dressings, I would look at them and weep."

She and Richard were as mystified by her infection as Saiman had been by the New York cases. But then in mid-July, while watching her husband's video of her delivery, Lore spotted something. The nurses who had taken Emmy aside to clean her up and check her were putting their fingers in Emmy's mouth to stimulate her sucking reflex before giving her to Diane to feed. Lore could see no evidence that they changed gloves or washed their hands first.

She contacted the hospital's quality-management department, asking for a discussion and offering to show them the tape. She met with them on July 22, and six days later received a letter. The hospital thanked her for the "troubling observations" she had made about the nurses, but declined to take any responsibility for her infection because it was a community strain. "From where did your drug-resistant infection arise? We cannot say with certainty," the letter said. "We are altogether too aware of the existence of MRSA and other resistant organisms in our communities."

Lore was slowly getting better. Her insurance company had finally agreed to pay for the linezolid, which arrived from a mail-order pharmacy in giant bottles of 140 pills at a time. First one drain and then the other was taken out of her breast, leaving shrunken puncture marks that puckered as they healed. On August 13, her infectious-disease physician told her that her blood work wasn't quite normal yet; she still had elevated levels of C-reactive protein, a blood component that indicates inflammation. But the doctor told her that if she could make it a month with no flare-ups, she could consider herself cured.

But on the morning of September 10, three days short, she woke up, rolled to her side, and felt a familiar sharp burning. A palm-sized red patch on the outside of her breast was spreading quickly over the skin and past the nipple. Near-incoherent with pain and panic, she called the doctor for help. It was possible the linezolid was not getting through her breast tissue, they told her. The best remedy would be vancomycin, but

201

it could only be given by IV. When she got to the doctor's office, the staff snaked a catheter through the veins of her arm so she could receive IV infusions at home. The infusions would be her responsibility to manage, though they booked phone time with a nurse to guide her.

Administering the infusions was nightmarish. Twice a day, while Richard managed the other kids, Lore and her nine-year-old son would struggle with the IV bags and the tubing that was meant to plug into the catheter port in her arm. The catheter needed to be flushed with a syringe to keep her blood from clotting in the end of it; neither of them knew how to do that. They were supposed to keep the tube and port sterile, but their fingers always seemed to slip and contaminate something. There were bubbles in the line, and they couldn't shake them free. Lore had been warned about air traveling to the heart or brain and causing a stroke, and it terrified her.

The situation terrified the rest of her family, too. Her younger son David was toilet-trained, but his day care providers told her he had been soiling his pants. Ian, the nine-year-old, was having nightmares. One day he said to his mother, "If you die, I'm going to commit suicide." Lore tried to reassure him, but in her heart she knew she wasn't convincing. She was frightened too, always aware that she was receiving what might be the last drug.

"You think, surely this can't be it," she said. "It's just an infection; surely there's something that will fix it. But we'd been through Zyvox, we'd been through vancomycin—what the hell else was there to try?"

After a week of struggle with the catheter, Lore's doctors took away the IV and put her back on linezolid. After four more weeks of drugs, she was declared cured, and left to put her life and family back together. And, not coincidentally, her career. After more than a decade writing about American medicine, her sympathy for its intricacies was permanently altered.

"The medical system completely destroyed my trust," she said. "It's greedy. It does nothing to keep people safe. It has no integrity anymore."

Across the country, groups of patients and parents were experiencing that same reaction. Institutions that they had trusted to keep them safe in their most unguarded moments were making them vulnerable to infections they had not foreseen. In 2003, at Texas Children's Hospital—the Houston institution where twelve-year-old Nick Johnson was treated and where the community strain's invasive attack on children was proven—eight very sick babies in the neonatal intensive care unit developed MRSA bacteremia. Six of the eight were infected with bacteria carrying SCC*mec* IV, the marker for the community strain. All six fell into septic shock, and three died.[6] Between 2001 and 2003, epidemiologists at the Children's Hospital of Pennsylvania in Philadelphia recorded rising levels of SCC*mec* IV staph among children who were hospitalized there.[7] In 2004, hospitals in Chicago and Los Angeles reported outbreaks of skin infections in their newborn nurseries. PFGE analysis revealed that both outbreaks were caused by USA 300, which was replacing MW2 as the dominant community strain.[8]

In 2005, a baby boy who had been born at Rady Children's Hospital in San Diego—the place where Bryce Smith would be treated and Carlos Don would die—came back to the hospital the same day that he was sent home with his parents. Immediately after birth, he'd been held in the hospital for five days because he wasn't gaining weight. Now his parents were bringing him back because he seemed feverish and his back and thighs looked puffy and flushed. It was good that they did. Their baby had necrotizing fasciitis, flesh-eating disease. It took two operations that denuded his back of flesh to save his life, and two sets of skin grafts to close his wounds. The baby's infection was caused by the community strain, USA300. But he could not have caught the bug in the community. He had only been in the world outside the hospital for mere hours.[9]

Around the same time, other hospitals reported that community-strain MRSA was also spreading among their adult patients. In 2004, at Grady Memorial Hospital in downtown Atlanta, one-fifth of MRSA bloodstream infections were caused by USA300.[10] At the veterans' medical center in Houston, the proportion was three-fifths.[11] At a community teaching hospital in St. Louis and at a university hospital in South Carolina, USA300 began causing significant numbers of infections in surgical incisions.[12,13] In Philadelphia, at Thomas Jefferson University Hospital, the community strain moved into the intensive care unit, and by 2006 it was causing 20 percent of the pneumonias that occurred in intubated patients on ventilators.[14]

Spreading the news of those outbreaks required that an individual institution have the enterprise to investigate the condition of its patients or the results in its microbiology lab, as well as the courage to announce the news at a medical meeting and the initiative to get it into a scientific journal. As with the emergence of hospital-acquired MRSA in the 1960s, and community-associated MRSA in the 1990s, there was no comprehensive national surveillance that could establish with confidence where or how fast this new epidemic was moving. Surveillance would be important not just for defining the new scope of MRSA's spread, but also because the characteristics of the cases might help explain why hospitals suddenly seemed to be friendly environments for the community strain. In large academic medical centers that conducted their own surveillance, there was no question that the community strain was finding fertile soil in hospitals. At Harbor-UCLA Medical Center south of Los Angeles, USA300 went from barely being present in hospitalized patients in 1999 to accounting for half of all hospital-acquired MRSA in 2004.[15] Stroger, Chicago's giant public hospital, saw a similar increase. Between 2003 and 2006, the overall rate at which MRSA caused bloodstream infections did not change—but by 2006, 50 percent of those infections were being caused by USA300, twice as many as three years before.[16]

For those who paid close attention, there had always been a bit of overlap between the hospital and community epidemics. CDC epidemiologists liked to emphasize a category that the agency called "hospital-acquired, community-onset," referring to patients who were infected during a hospital stay but didn't develop symptoms until they returned home. Once these strains left the hospital, they sometimes spread, and people who had no known contact with health care might nonetheless be revealed on microbiological analysis to be infected with USA100 or one of the other hospital strains.[17] The CDC could not say with precision how much hospital MRSA escaped in this manner, because much of the agency's surveillance relied not on microbiology but on risk factors, such as prior hospitalization or current dialysis, to define the categories into which MRSA cases fell. Researchers in community MRSA disagreed with the CDC's categories. With USA300 so prevalent, they felt that past contact with the health care system had no predictive value for future infection; someone might have brought a hospital strain home with them, but it was equally possible they might have been colonized with a community strain after they left the hospital, or even before they went in.[18]

As Saiman had foreseen, the converging of the epidemics meant that fifty years of medical terminology was sliding out of date. "Hospital strain" and "community strain" had once been useful designations because they were shorthand for many differences—not only the places where transmission occurred and the conditions that made spread more likely there, but the resistance profiles the strains exhibited and the kinds of illnesses they caused. Chip Chambers and Francoise Perdreau-

Remington had revealed in 2006 that the USA300 community strain in San Francisco and Boston was becoming resistant to more drugs. Now the newest hospital outbreaks were demonstrating that USA300 could create the same surgical and bloodstream infections that USA100 had traditionally caused, while attacking new groups of patients who had not previously been considered at risk.

With the ground rapidly shifting beneath them, hospital epidemiologists were paralyzed by a vital question: if USA300 moved into their hospitals, how could they detect its emergence in time to block its spread? Hospitals were already struggling with whether to implement active surveillance. Though search and destroy methods weren't cheap, or easy to incorporate into an institution's work flow, they only required two uncomplicated steps: swab the nose and culture or rapid-test the swab. But as cases like Mollie Logan's daughter Isabella had proven, and recent research had confirmed, USA300 did not necessarily hang out in the nose.[19] It was carried on other sites in the body, the vagina and rectum in particular, that would be much more challenging to swab during a quick admission screening. Nevertheless, if USA300 was not identified as it entered hospitals, infection control efforts would be undermined.

There was an additional concern. Microbiologically, the thing that had distinguished the community strains from the hospital strains was USA300's production of PVL and other toxins. Those toxins gave community staph the power to liquefy lung tissue, create toxic shock–like syndromes, and start fast-expanding infections on unbroken skin. By and large, the toxins were not present in the hospital strains; those strains' power came instead from the frailty of the hospital patients they attacked. USA300's move into hospitals would bring it under much greater antibiotic pressure than existed on the outside. If such a move in turn brought about an exchange of genetic information, marrying the multidrug resistance of the hospital strain with the transmissibility and virulence of the community strain, the result would be a superbug indeed.

The few infectious-disease specialists who were alert enough to MRSA's behavior to understand this danger began watching their local epidemics for the emergence of such a bug. In the summer of 2007, the University of Chicago hospitals, home of Robert Daum's research team, believed they had found it.

At the University of Chicago's Comer Hospital, the outbreak started with a premature baby so tiny that doctors struggled to find adequate medical terminology to describe her. Preemies' ability to survive is determined by their gestational age—the number of weeks since conception, up to birth at forty weeks—and their weight when they leave the womb. The earlier the gestational age and the lower the birth weight, the greater the risk of infection, because the immune system has yet to finish maturing and the lungs are incomplete. Neonatologists consider a child "low birth weight" at 2,500 grams, "very low birth weight" at 1,500 grams, and "extremely low birth weight" at 1,000 grams. The Comer baby was born at twenty-five weeks, weighing 671 grams or 1 pound, 7 ounces—about as high-risk as an infant can possibly be.[20]

She was immediately admitted to Comer Hospital's neonatal intensive care unit, intubated to get as much oxygen into her as possible, and given a central line, an implanted catheter that ends up near the heart, to make it easier to draw blood and give medications. She was started on a beta-lactam antibiotic that could target multiple families of bacteria, another common measure because preemies are so vulnerable to infection. When she was a month old, her health care workers started noticing changes in her blood work and heart rate, and that her ventilator settings needed more tinkering than usual. Preemies are already so sick that diagnosing any additional infection is a challenge, but within a few days her doctors had confirmation that something was seriously wrong. A CT scan showed a developing pneumatocele, a walled-off, air-filled space inside her lung. The space existed because necrotizing pneumonia had eaten away her lung tissue, opening a hole into an airway so that air rushed into the empty space and inflated it when the ventilator breathed for her. The pneumonia was an incontrovertible sign that the baby had MRSA, and in confirmation the bacterium was isolated from her blood. On analysis, the lab found it was resistant to beta-lactams and erythromycin, but susceptible to clindamycin and trimethoprim-sulfamethoxazole. The strain carried SCC*mec* IV and made PVL. The baby, who'd never left the hospital, had USA300, the community strain.

In a child so fragile, USA300 was vicious. Critically ill and bacteremic, she was put on vancomycin. It took six weeks of the drug to quell the infection. She remained in the NICU for more than 100 days. A month after her diagnosis, while she was still being treated, a second baby at Comer was diagnosed with necrotizing pneumonia. This time the baby was a slightly more robust preemie, who had been born at twenty-five weeks weighing 900 grams or almost 2 pounds, and was now seventy days old. Twelve days after that, a third, brand-new preemie—born just eight days earlier at 912 grams and twenty-seven weeks—developed the same infection. Like the first child, the second and third babies had MRSA in their breathing tubes and in their blood, and huge air-filled cavities in their lungs. Both were sick, but one was severely ill, with depressed blood pressure, kidney failure, and the beginnings of disseminated intravascular coagulation, the abnormal clotting in small blood vessels that signals septic shock. Both of them were put on intravenous clindamycin. After four weeks they were mostly recovered from the immediate infections; still, each remained in the NICU for more than 100 days.

By the second case, the NICU realized it had an outbreak on its hands, and began checking all of its babies for the bacteria. They found an additional three that were carrying USA300 but were not ill. One of the colonized was a twin of the sickest preemie, but she did not develop the severe disease that had so undermined her sister.

The Chicago team's investigation could not tell them how the community strain had arrived in the NICU. Perhaps it had come in with a nurse or technician, and passed from that person to a baby and then back to a health care worker and on. Perhaps one baby had received the strain from its mother at birth—or perhaps, given USA300's ubiquity, all of them had. Whatever route USA300 took into the NICU, the team now understood that the strain could penetrate even the most controlled and carefully monitored corners of the hospital and cause disease there. Thankfully, by good luck or good infection control, USA300 had not spread very far. But it had long-term effects: the first baby to fall ill, the one with the lowest birth weight, never fully recovered from the necrotizing pneumonia. She made it home, but six weeks later returned to the hospital, where the damaged part of her lung was removed. She died of respiratory failure when she was eight months old. And even with tightened infection control and heightened awareness, USA300 struck again. In November 2008, sixteen months after the three preemies fell ill, the same strain reappeared in the Chicago NICU, in a fourth preemie. Once again, it caused necrotizing pneumonia. In a short time, the baby died.

Daum's Chicago team had spent decades researching MRSA, and more than a decade on the community strain alone. By now they were intimately familiar with staph's almost willful ability to find new victims, new ways to express its virulence, and new strategies for evading every attempt at antibiotic control. They concluded that there was really only one thing that would end the long struggle with the bug: finding a vaccine.

A vaccine could not curb MRSA's ability to develop resistance to every drug brought against it—at least, not directly. But if it could reduce the amount of staph infection that occurred, it could stall or diminish resistance, because it would reduce the amount of antibiotics used in MRSA patients, and thus reduce the selective pressure the drugs placed on the bug.[21]

There were precedents for choosing prevention over endlessly trumped attempts at a cure. Twice in the past two decades, medicine had successfully developed and deployed vaccines against two prevalent bacteria that were developing resistance and causing increasing amounts of disease: pneumococcus and *Haemophilus influenzae* type B, usually known as Hib. The Hib vaccine vastly reduced rates of bacterial meningitis, and also the degree to which children were asymptomatically colonized with the bacterium. The Prevnar vaccine had a similar effect on the incidence of pneumococcal pneumonia, though it did not reduce colonization in the same way.[22-24] In a much larger attempt at prevention, researchers in the United States and elsewhere had been seeking an AIDS vaccine almost since the start of the HIV epidemic. After more than twenty years, the search was still unsuccessful, but it continued, with federal government support of hundreds of millions of dollars each year—in any year, more for that one vaccine effort than was given to all antimicrobial research combined.[25] And MRSA, as the CDC had demonstrated, caused more U.S. deaths each year than AIDS did.[26] In 2008, CDC researchers estimated that achieving a vaccine could reduce invasive MRSA cases by 12 percent to 47 percent, preventing up to 49,940 infections each year.[27]

But achieving a MRSA vaccine poses unique challenges. One is deciding to whom to give it: those at the highest risk of hospital infection, such as dialysis recipients or patients about to get a prosthetic joint or implanted device? The otherwise healthy who are at high risk of community infection, such as school kids and prison guards? People experiencing recurrent infections? The elderly entering nursing homes? Premature babies admitted to a NICU?

Equally challenging is determining which aspect of staph's enormous armamentarium to target. A vaccine could, with varying degrees of difficulty, attack the attachment proteins that allow the bacterium to bind to a cell, the host-defense factors that hide and protect it from the immune system, or the wide array of toxins such as PVL and the superantigens that contribute to MRSA's virulence.[28] A third hurdle is the most devastating of all. Vaccines rely for their effectiveness on provoking an immune response in the body. Even with decades of research, there is no agreement that lasting immunity to staph exists—a vital question to answer before any vaccine work begins, since measuring a change in immunity is an essential regulatory test for proving the worth of a candidate vaccine.[29]

Despite the challenges, several serious attempts at a staph vaccine have been made, though none have succeeded. StaphVAX, made by Nabi Biopharmaceuticals, looked promising when the results of its early trials were published in the 1990s. The formula was analogous to that of Prevnar, the pneumococcus vaccine, in that it was based on molecules from an outer casing that prevents bacteria from being engulfed by white blood cells. In a Phase III trial in dialysis patients, StaphVAX scored positive results. It produced a measurable degree of immunity in recipients, though that immunity faded after forty weeks.[30] But when the company repeated the trial with a larger number of patients and a booster dose, its results were no better than the placebo used against it, and Nabi suspended the program in 2005.[31] A second vaccine attempt by a company called Inhibitex, using a formula they dubbed Veronate, aimed for what is called passive immunization—meaning that, rather than stimulating the body to make antibodies, it delivers already-formed antibodies harvested from blood donors. Passive immunization would be especially useful for those who are immediately threatened with staph infection and have no time to make antibodies of their own, such as preemies and patients undergoing emergency surgery. Veronate had also done well in early studies, but failed in its final trial, delivering a result that was no better than placebo.[32]

Both trials were considered disappointments. Several vaccine-development programs persisted, however, including one based around a reformulation of StaphVAX, and another developing a novel formula that targeted cell membrane protein, which was the work of pharmaceutical giant Merck & Co.[33] In 2008, researchers reported interim success—in an animal trial only—with a third new approach. They protected mice against staph pneumonia using a vaccine that induced antibody response to one of staph's virulent toxins, one that had already been implicated in human pneumonia as well.[34]

In his paper-choked office in University of Chicago's old Wyler building—long since converted to offices, but once the hospital where he and Betsy Herold had stopped to exchange their impressions of the first community infections—Robert Daum brooded over MRSA's many costs. Millions of doctor and emergency room visits; hundreds of thousands of hospitalizations; longer, costlier hospital stays; and tens of thousands of deaths. For years he had witnessed the havoc wreaked by the bacterium he had come to regard, against his scientific instincts, almost as a personal opponent. Daum wondered whether man's longtime companion and killer could ever be controlled.

"A vaccine is what we need to do," he said. "We know what the attack rates are. We know how much MRSA invasive disease there is. We have every argument we need to support this. What we need is the funding, and the political will that would make it possible. I truly believe it is the only way."

EPILOGUE

Tony Love was discharged from La Rabida Children's Rehabilitation Hospital in February 2008. According to Comer Children's Hospital, he and Clarissa missed several follow-up appointments that spring and summer; the hospital closed their file in June and has not heard from the family since.

In August 2009, Jay Sorensen had his eighth surgery, to repair four new hernias. Once again, he developed MRSA; after a week in the hospital, he was sent home once again on IV vancomycin, plus several other drugs. His doctors warn him that his abdominal muscles have been so weakened that he can expect periodic surgeries for the rest of his life. In fall 2009, he was hired by WCBS-FM in New York City.

In November 2005, Everly Macario and James Sparrow had a second son, Dylan. When Dylan reached eighteen months, passing the age that Simon was at his death, Macario went back to work. She joined the pediatric infectious diseases team at the University of Chicago hospitals, the institution where Simon died, working for Dr. Robert Daum, who had figured out the cause of his death.

Steve McNees was eventually treated with a combination of trimethoprim-sulfamethoxazole and rifampin. He had one more recurrence of an abscess in late 2007, but since then has been MRSA-free.

Mollie Logan gave birth to her and Brian's second child, Ian, in April 2006. Her doctors checked her repeatedly for a month before the birth, planning to administer antibiotics if they found her colonized nasally or

211

vaginally, but she was always clear. Ian has never contracted MRSA. Isabella remains symptom-free.

One month after Carlos Don died, the two medical societies that write the guidelines for pneumonia treatment changed their recommendations. They advised everyday physicians to assume that pneumonia cases are caused by MRSA until test results prove the assumption to be wrong, and to prescribe a drug that is active against MRSA so that the pneumonia's destruction can be halted in time.[1] In 2008, in part because of concern about MRSA deaths after flu infections, the federal Advisory Committee on Immunization Practices recommended that all children and teenagers, up to the age of eighteen, receive flu shots.[2] And, about eighteen months after Carlos died, the Infectious Diseases Society of America and several other specialty societies agreed to write the first guidelines that would advise doctors how to treat community pneumonia and MRSA pneumonia in children.

Scott and Katie Smith and Amber and Carlos Don III both filed suits against the physicians who were the first to see Bryce and Carlos. In both cases, the physicians denied all liability. In 2008, the Smiths' suit was dismissed without prejudice. In October 2009, a jury found negligence against one of the physicians who first saw Carlos, but deadlocked on a second charge, and the case ended in a mistrial.

Bucks County, Pennsylvania, and its codefendants appealed the $1.2 million award to Kevin Keller and Benjamin Martin. In December 2006, the award was affirmed by the Third Circuit Court of Appeals.

After 2003, Diane Lore had no recurrences of MRSA. Three years later, when she needed surgery, she opted to have it done in Mexico instead.

APPENDIX

Methicillin-resistant *S. aureus*, MRSA, has been resistant to penicillin for decades. It is also resistant to methicillin and the other semi-synthetic penicillins, which were based on the original, natural penicillin molecule. What all of these have in common is a four-cornered arrangement of atoms at the center of the drug molecule, called the *beta-lactam ring*. That ring first appeared in natural penicillin and was retained by pharmaceutical chemists when they created methicillin and its close relatives in the 1960s. When the bacterium learned how to disable the beta-lactam ring (in two different maneuvers, one for penicillin and one for the semi-synthetics), it made itself resistant to all the penicillin-based drugs.

In addition to methicillin, which has passed out of common use, the semi-synthetics include oxacillin (used in laboratories to test for methicillin resistance), nafcillin, cloxacillin, dicloxacillin, and flucloxacillin. Ampicillin and amoxicillin, semi-synthetic penicillins that were developed more recently, also incorporate the beta-lactam ring. Amoxicillin is a component of the combination drug Augmentin, which contains a second compound, clavulanic acid.

However, a number of other classes of common antibiotics also incorporate the ring in their chemical structure. Collectively, they are called *beta-lactam antibiotics* and are the most commonly prescribed drugs in medicine. And to one degree or another, depending on the genetics of the particular strain, MRSA is resistant to them (though not all of them would normally be prescribed for MRSA, since some are approved only to treat other organisms).

Drugs possessing the beta-lactam ring include:

The *cephalosporins*, which have been around since the late 1940s. There have been several "generations" of them. Drugs in the earlier generations of cephalosporins include cephalexin (Keflex), cephalothin, cefazolin, cefaclor, cefotaxime, and ceftriaxone. There is currently a fifth generation of cephalosporins on which researchers pin great hope, exemplified by the drug ceftobiprole.

The *carbapenems*, available since the 1970s, which include imipenem and meropenem among others.

The *monobactams*, developed in the 1980s, which include aztreonam.

Resistance to beta-lactam antibiotics is the hallmark of all MRSA strains. Many strains have also picked up resistance to additional drug families that do not possess the beta-lactam ring; they did this by incorporating gene segments that were acquired from other bacteria into staph's genetic code.

Multi-drug resistance was originally the hallmark of MRSA strains that circulated in hospitals. By the 1990s, those strains were resistant not only to the beta-lactams, but to at least some drugs in these families as well: *macrolides* (erythromycin); *lincosamides* (clindamycin); *aminoglycosides* (gentamicin); *fluoroquinolones* (ciprofloxacin); and *tetracyclines*.

Strains of MRSA that caused community infections were originally resistant only to the beta-lactams, but many of them have now picked up at least some of the resistance factors that were originally the province of hospitals strains, including to erythromycin, clindamycin, tetracycline, and sulfa drugs (part of the combinations Bactrim and Septra). Because community MRSA in particular is so genetically diverse, the resistance factors it possesses are not uniform—so one infection may be susceptible to drugs that another infection may be unaffected by.

MRSA has also been acquiring resistance to mupirocin (Bactroban), the active ingredient in the ointment used to decolonize MRSA carriers. It has developed several types of resistance to vancomycin, a *glycopeptide*, which for decades was the MRSA drug of last resort, and is beginning to exhibit resistance to drugs in newly developed families that were intended to replace vancomycin: linezolid (Zyvox), an *oxazolidinone*, and daptomycin (marketed as Cubicin), a *lipopeptide*. In September 2009, the FDA approved a long-awaited new drug in another new class (*lipoglycopeptides*, partially modeled on vancomycin) called telavancin

(Vibativ). But telavancin was approved only for serious skin and "skin structure" infections, not for the most serious invasive syndromes such as osteomyelitis, bacteremia, or endocarditis.

Determining whether or not a MRSA infection is susceptible or resistant to a particular drug is an essential part of caring for MRSA infections, no matter where they occur or how serious they are. If you receive a prescription to treat MRSA, insist that your doctor discuss with you what names the drug is sold under, what drug family it belongs to—and most important, what reasons your doctor has for believing that drug will be effective.

NOTE ON SOURCES
AND ACKNOWLEDGMENTS

This book is based on approximately two hundred interviews with physicians, scientists, policy makers, victims who have survived MRSA, and friends and family members of those who did not survive or could not speak for themselves. I am profoundly grateful to all of them for their thoughtfulness, courage, and patience with my endless questions. It is also based on my review and analysis of more than 1,100 scientific papers and conference presentations, and on approximately two hundred newspaper and magazine articles dating back to the 1940s.

I conducted all the interviews on which this book is based; about half were done in person and the rest by phone. The interviews were recorded and the majority transcribed by professional transcribers, though I transcribed a small proportion of them myself.

I endeavored to check accounts of events against medical records, legal documents, family papers and diaries, maps and photographs, and contemporaneous news coverage. Most episodes were reviewed by at least one major participant in the events described.

This book's genesis was a story about community-associated MRSA that I wrote for *Self* magazine in February 2007; without the vision of Sara Austin, news and health features director, the story would never have seen print, and I am grateful to continue to work with her. It expanded into a book proposal under the unfailing guidance of my brilliant agent Susan Raihofer of the David Black Literary Agency. It became a book thanks to the stewardship and acumen of my editor Emily Loose (and her assistant Maura O'Brien) at Free Press, as well as editor in chief Dominick Anfuso and publisher Martha Levin. I owe special thanks to

Elizabeth Stein, editor of my first book, *Beating Back the Devil*, who saw the promise in *Superbug* and acquired it for Free Press.

When I began work on the book, I was a media fellow with the Henry J. Kaiser Family Foundation, working on a project on emergency room chaos in which I saw far too much MRSA not to be intrigued by it. I thank Penny Duckham, Deirdre Graham, Matt James, and Drew Altman for their care and support for many classes of journalists. As I finished it, I was named an Ochberg Fellow of the Dart Center for Journalism and Trauma at Columbia University, and I thank Bruce Shapiro, Dorie Griggs, Frank Ochberg, MD, and my fellow fellows at Dart for helping me understand how to tell the stories of MRSA victims with dignity and truth.

I thank my first readers Barth Anderson, Frances Katz, and Quinton and Jean Gregor; and Anne Edwards, MD, who read the revised version. Thanks also to my Kaiser co-fellows Joanne Kenen, who read the first chapter early on, and Howard Gleckman, who simultaneously was writing his excellent book *Caring for Our Parents* and commiserated at all the right moments.

For most of the time that I was reporting this book, I maintained a blog, also called *Superbug*, as online notes and digital whiteboard. It has attracted a broad community of readers. I am grateful for every one of them, and for the support of fellow bloggers Michael Coston of *Avian Flu Diary*, Crawford Kilian of *H5N1*, Scott McPherson, Liz Borkowski of *The Pump Handle*, and the determinedly pseudonymous "Revere" of *Effect Measure*.

I extend humble thanks to my colleagues at CIDRAP, the Center for Infectious Disease Research and Policy at the University of Minnesota, who tolerated my frequent absences and unnerving obsessions: Michael T. Osterholm, PhD; Kristine Moore, MD; Jim Wappes, Robert Roos, and Lisa Schnirring; Kathleen Kimball-Baker; Nicholas Kelley, who shared his enthusiasm for phages; and Paul Mamula, PhD, who scouted literature for me and helped me understand obscure references. The reference staff of the University of Minnesota's Biomedical Library were always courteous, swift, and smart, and they helped me a great deal.

I benefited from the sterling work of several groups and think tanks who have done deep explorations of subjects I touch on only briefly, including Coilin Nunan at the Soil Association; Ramanan Laxminarayan, PhD, and his colleagues at the Extending the Cure project of Resources for the Future; Donald Berwick, MD, Donald Goldmann,

MD, Frances Griffin, Joseph McCannon and Madge Kaplan of the Institute for Healthcare Improvement; Lisa McGiffert of Consumer Reports; the Pew Commission on Industrial Farm Animal Production; the Center for a Livable Future at Johns Hopkins; and the Union of Concerned Scientists. Of course, Dr. Stuart Levy of Tufts University and the Alliance for the Prudent Use of Antibiotics is the dean of all this work, and I acknowledge my debt to him and the many researchers he inspired.

James Sliwa of the American Society for Microbiology for years has been a thorough, accurate, and incredibly good-humored guide through the mysteries of medical meetings, as has Steven Baragona, formerly of at the Infectious Diseases Society of America.

It is not possible to do a book this kaleidoscopic without receiving help from many people, and so there are many names whom I must mention in thanks.

In Chicago, the entire team of the University of Chicago's MRSA Research Center was very kind to me: Robert S. Daum, MD, Susan Boyle-Vavra, PhD, Everly Macario, SciD, Michael Z. David, MD, PhD, Susan Crawford, MD, Christopher Montgomery, MD, Ben Yoon, and Lakesha Lloyd; and also Robert Bielski, MD, Stephen Weber, MD, and John Easton.

In San Francisco, the staff of San Francisco General and the University of California, San Francisco, were equally helpful: Henry "Chip" Chambers, MD, Francoise Perdreau-Remington, PhD, Adam Hersh, MD, Binh An Diep, PhD, and Kristen Bole; elsewhere in the city, Rick Loftus, MD, and Steve Gray. Moving south, in Los Angeles I benefited from the knowledge of Elizabeth Bancroft, MD, Nolan Lee, MD, Gregory J. Moran, MD, John Edwards, Jr., MD, and Loren G. Miller, MD.

In San Diego, John Bradley, MD, and Bradley Peterson, MD, of Rady Children's Hospital went far beyond courtesy in helping me understand their cases and challenges. On the veterinary side, so did Don Janssen, DVM, Nadine Lamberski, DVM, and Jeff Andrews of the San Diego Zoo Wild Animal Park.

In Georgia, I was grateful to have the help of Nicole Coffin and David Daigle, Rachel Gorwitz, MD, Jeffrey Hageman, Fred Tenover, PhD (now with Cepheid), Timothy Naimi, MD, Denise Cardo, MD, Joan Brunkard, PhD, David Sugerman, MD, Roberta Carey, PhD, Brandi Limbago, PhD, Lonnie King, DVM, and Jennifer Wright, DVM, at the Centers for Disease Control and Prevention; Susan Sanchez, PhD, at the University of Georgia; at Emory University School of Medicine, Arthur Kellermann,

MD, Katherine Heilpern, MD, Henry Blumberg, MD, Leon Haley, MD, Michael Huey, MD, Andre Nahmias, MD, and Holly Korschun; and Tom Keating, PhD, of Project CLEAN.

In Texas, many people kindly helped me understand the foreign world of high school football: David Edell, Cathy Supak, Mike Carroll, and Marilyn Felkner, MD. Thanks also to Heinz Eichenwald, MD, for reviewing the long-ago days of staph 80/81, and to Mari Nicholson-Preuss for sharing her research on the Jefferson Davis Hospital outbreak.

Many others coached me through my encounters with the closed world of the correctional system: Gregory Belzley, Martha Sperling, Anita Alberts, Nate Wenstrup, Natalie Leonard, Carrie Takahata, and Eric Balaban of the American Civil Liberties Union; John Clarke, MD, Luis Rivera, MD, Timothy Faloon and Rhonda Moore; Charles Alexander, Jr., Gary Benson, David White, Chris Pearson; Pierre Tattevin, MD, and Jason Farley, PhD; and Elizabeth Cumming, Katie Schwartzmann, and Samuel Gore, MD.

Many more people helped me understand the world of veterinary medicine and the too-often ignored links between animal health and our own: Tara C. Smith, PhD, Michael Male, DVM, and Abby Harper at the University of Iowa; J. Scott Weese, DVM, at the University of Guelph Ontario Veterinary College; Carlo Vitale, MD, in San Francisco; Andreas Voss, MD, PhD, and Jan Kluytmans, MD, PhD, in the Netherlands; and Pat Gardiner in the United Kingdom.

In the realm of infection control, I could not have had better guides than Barry M. Farr, MD, and William Jarvis, MD. I often turn for epidemiological understanding to William Schaffner, MD, of Vanderbilt University, who has helped me more times than I can accurately count. In Minneapolis, Jeff Bender, DVM, B. J. Anderson, MD, John Odom, MD, Marjorie Hogan, MD, Kirk Smith, DVM, and dean of staph researchers Patrick Schlievert, PhD.

At Novant Health, I thank Paul M. Wiles, Thomas N. Zweng, MD, Stephen L. Wallenhaupt, MD, James W. Lederer, Jr., MD, Suzie Rakyta, RN, Sandy Cox, RN, Patty Deltry, RN, and Kati Everett. I also benefited from great courtesy at Pitt County Memorial Hospital in Greenville, North Carolina, especially from Keith M. Ramsey, MD, Kathy Cochran, RN, and Delores Nobles.

At Cubist Pharmaceuticals, Barry Eisenstein, MD, Richard Baltz, PhD, Grace Thorne, PhD, and Jared Silverman, PhD, were gracious, thorough, and precise. I am especially grateful to Kevin Tally, MD, for sharing with me the story of the life, accomplishments, and death of his father Frank Tally, MD. Elsewhere in Boston, I thank Joann Lindenmayer, DVM, Robert Moellering, MD, and Jeffrey Linder, MD. For additional help in understanding pharmaceutical regulation, I thank John Powers, MD, formerly of the Food and Drug Administration; and for the intricacies of drug and vaccine development, Ron Najafi, PhD, of Nova Bay Pharmaceuticals, Mary Beth Minyard of the Southern Research Institute, and especially Henry Shinefield, MD. Lisa Bayne, corporate archivist for Eli Lilly and Company, helped me excavate the murky early history of daptomycin.

Also, in no order save alphabetical, because all were gracious and thoughtful and I can think of no way to rank their efforts: Allison Agwu, MD; Steven J. Barenkamp, MD; Gonzalo Bearman, MD; Benjamin Estrada, MD; Eugene L. Green; Brenda Hollier; Christoph U. Lehmann, MD; Nkuchia Mikinatha, DrPH; Michael Nagy, MD; Louis Rice, MD; Lisa Saiman, MD; and Torrance Williams.

I have been humbled by the readiness of MRSA victims and their families to share what are often very difficult and painful stories. Among them are: Rose Aliberti; Kepa Askenasy; James Bell; Michael Bennett; Laura Chen Davies, Robin Cook; Stephanie Crowell; Amber Don; Eva Ferguson and Beverly and Ernie Dieringer; Nick and Janet Johnson; Kevin Keller; Mollie and Brian Logan; Diane Lore and Richard Ross; Clarissa and Tony Love; Christina Marzullo; Jane McGuinn, Molly Donahue, Charlotte McGuinn Freeman; Steve, Sue, and Pat McNees; Karen Moser; Carol and Ty Moss; Victoria Nahum; Melissa Quintana; Debbie Russ; Danielle Sheffler; Scott and Katie Smith; Andrea and Jay Sorensen; Jeanine Thomas; Suzanne Turpin Upton; and Tom Wolff. I am particularly grateful to Christine Fusco and Lourraine Stamets, who shared with me the painful stories of the deaths from MRSA of their husbands.

After I wrote about MRSA for Self and Health magazines, dozens more MRSA victims contacted me to tell me their stories, but asked me to keep their experiences private. I respect their wishes and am grateful for their trust.

I benefited to an extraordinary extent from the generosity of friends and family who both put me up and put up with me, mixed me cocktails, and listened to stories that made them look around furtively for Purell: in Atlanta, Samara Cummins, Krista Reese, and Rich Eldredge. In Washington, D.C., Elizabeth Robelen and Carter Reardon. In Annapolis, Lt. Col. Robert J. McKenna, Cerise McKenna Vablais, PhD, and my nephew

Zach McKenna, who kindly loaned me not only his bunk bed and stuffed animals, but also his light-saber to defend myself against stray pathogens. In New York City, my uncle Rev. Robert E. Lauder, my family's much more distinguished author, who supplied Italian food, emotional support, and emergency funds. In Houston, my father John J. McKenna, who not only explained the back roads to Stafford, but unraveled the mysteries of the New York City subway system, noting that the trains Beverly Dieringer took to the New York Hospital were the same ones he took to college at Columbia University back before the track names changed. In Minneapolis, Heather Cummings. In California, Matthew McKenna and Darla Albright in Los Angeles, David Tuller in San Francisco, and Randy Dotinga in San Diego. My sister Elizabeth McKenna has never had to put me up on a research trip, because, mysteriously, nothing disease-related seems to happen in Kalamazoo, but I thank her nonetheless and just in case.

And before, behind, and above all else, my husband Loren D. Bolstridge III, who did my laundry, booked my flights, made me mix-tapes, brought me tea when I needed it and single malt when I needed it more, and patiently endured far too many weekends when I opted for my computer's company over his. I could not have done this without him, and I would not have wanted to.

NOTES

1. THE FIRST ALERT

1. The account of Tony Love's experience is based on interviews with Tony and his mother Clarissa Love at the University of Chicago on December 5, 2007; and on interviews with members of his treatment team, including Dr. Robert Daum (December 3 and 4, 2007, August 22, 2008, December 12, 2008, and June 18, 2009), Dr. Robert Bielski (February 28, 2008), and Dr. Christopher Montgomery (December 4, 2007, and January 3, 2008).
2. In addition to the cited articles, the history of the University of Chicago team's engagement with MRSA is assembled from interviews with Dr. Robert S. Daum (June 24, 2006, December 3, 2007, August 22, 2008, and June 18, 2009) and Susan Boyle-Vavra, PhD (December 12, 2007, December 19, 2008, and March 16, 2009).
3. Kuehnert MJ, Kruszon-Moran D, Hill HA, et al. Prevalence of Staphylococcus aureus nasal colonization in the United States, 2001–2002. *J Infect Dis* 2006;193:172–79.
4. Crossley KB, Archer G. *The Staphylococci in Human Disease* (New York: Churchill Livingstone, 1997).
5. Kuroda M, Ohta T, Uchiyama I, et al. Whole genome sequencing of meticillin-resistant Staphylococcus aureus. *Lancet* 2001;357:1225–40.
6. Klein E, Smith DL, Laxminarayan R. Hospitalizations and deaths caused by methicillin-resistant Staphylococcus aureus, United States, 1999–2005. *Emerg Infect Dis* 2007;13:1840–46.
7. Abraham E, Chain E. An enzyme from bacteria able to destroy penicillin. *Nature* 1940;146:837.
8. Jevons MP. Celbenin-resistant staphylococci. *Br Med J* 1961:124–25.
9. Panlilio AL, Culver DH, Gaynes RP, et al. Methicillin-resistant Staphylococcus aureus in U.S. hospitals, 1975–1991. *Infect Control Hosp Epidemiol* 1992;13:582–86.
10. Herold BC, Immergluck LC, Maranan MC, et al. Community-acquired methicillin-resistant Staphylococcus aureus in children with no identified predisposing risk. *JAMA* 1998;279:593–98.
11. Cafferkey MT. *Methicillin-resistant Staphylococcus aureus: clinical management and laboratory aspects* (New York: M. Dekker, 1992).
12. Boyce JM. "Are the epidemiology and microbiology of methicillin-resistant Staphylococcus aureus changing?" *JAMA* 1998;279:623–24.

13. Saravolatz LD, Pohlod DJ, Arking LM. Community-acquired methicillin-resistant Staphylococcus aureus infections: a new source for nosocomial outbreaks. *Ann Intern Med* 1982;97:325–29.

14. Schwartz DC, Cantor CR. Separation of yeast chromosome-sized DNAs by pulsed field gradient gel electrophoresis. *Cell* 1984;37:67–75.

15. Berman D, Holzman R, Della-Latta P. Epidemiology of community acquired methicillin-resistent S. aureus in intravenous drug abusers. *26th Interscience Conference on Antimicrobial Agents and Chemotherapy;* 1986 September 28–October 1; New Orleans; 1986.

16. Berman DS, Eisner W, Kreiswirth B. Community-acquired methicillin-resistant Staphylococcus aureus infection. *N Engl J Med* 1993;329:1896.

17. Pate KR, Nolan RL, Bannerman TL, Feldman S. Methicillin-resistant Staphylococcus aureus in the community. *Lancet* 1995;346:978.

18. Levine DP. Vancomycin: a history. *Clin Infect Dis* 2006;42 Suppl 1:S5–12.

19. Micek ST. Alternatives to vancomycin for the treatment of methicillin-resistant Staphylococcus aureus infections. *Clin Infect Dis* 2007;45 Suppl 3:S184–90.

20. Moellering RC, Jr. Vancomycin: a 50-year reassessment. *Clin Infect Dis* 2006;42 Suppl 1:S3–4.

21. Sievert DM, Rudrik JT, Patel JB, McDonald LC, Wilkins MJ, Hageman JC. Vancomycin-resistant Staphylococcus aureus in the United States, 2002–2006. *Clin Infect Dis* 2008;46:668–74.

22. Tenover FC, Moellering RC, Jr. The rationale for revising the Clinical and Laboratory Standards Institute vancomycin minimal inhibitory concentration interpretive criteria for Staphylococcus aureus. *Clin Infect Dis* 2007;44:1208–15.

23. Infectious Diseases Society of America. *Bad Bugs, No Drugs.* Washington, D.C., July 2004.

24. Daum RS. Clinical practice. Skin and soft-tissue infections caused by methicillin-resistant Staphylococcus aureus. *N Engl J Med* 2007;357:380–90.

25. Tsiodras S, Gold HS, Sakoulas G, et al. Linezolid resistance in a clinical isolate of Staphylococcus aureus. *Lancet* 2001;358:207–8.

26. Moellering RC, Jr. Current treatment options for community-acquired methicillin-resistant Staphylococcus aureus infection. *Clin Infect Dis* 2008;46:1032–37.

27. Gorak EJ, Yamada SM, Brown JD. Community-acquired methicillin-resistant Staphylococcus aureus in hospitalized adults and children without known risk factors. *Clin Infect Dis* 1999;29:797–800.

28. Centers for Disease Control and Prevention. Methicillin-resistant Staphylococcus aureus skin or soft tissue infections in a state prison—Mississippi, 2000. *MWWR, Morb Mortal Wkly Rep* 2001;50:919–22.

29. Fergie JE, Purcell K. Community-acquired methicillin-resistant Staphylococcus aureus infections in south Texas children. *Pediatr Infect Dis J* 2001;20:860–63.

30. Groom AV, Wolsey DH, Naimi TS, et al. Community-acquired methicillin-resistant Staphylococcus aureus in a rural American Indian community. *JAMA* 2001;286:1201–5.

31. Centers for Disease Control and Prevention. Outbreaks of community-associated methicillin-resistant Staphylococcus aureus skin infections—Los Angeles County, California, 2002–2003. *MWWR, Morb Mortal Wkly Rep* 2003;52:88.

32. Centers for Disease Control and Prevention. Four pediatric deaths from commu-

nity-acquired methicillin-resistant Staphylococcus aureus—Minnesota and North Dakota, 1997–1999. *MWWR, Morb Mortal Wkly Rep* 1999;48:707–10.

33. Interview with Timothy S. Naimi, MD, February 7, 2008
34. Mongkolrattanothai K, Boyle S, Kahana MD, Daum RS. Severe Staphylococcus aureus infections caused by clonally related community-acquired methicillin-susceptible and methicillin-resistant isolates. *Clin Infect Dis* 2003;37:1050–58.
35. Adem PV, Montgomery CP, Husain AN, et al. Staphylococcus aureus sepsis and the Waterhouse-Friderichsen syndrome in children. *N Engl J Med* 2005;353:1245–51.
36. Purdy JB. An enemy in our midst. *Medicine on the Midway* 2006:26–33.
37. Daum RS, Ito T, Hiramatsu K, et al. A novel methicillin-resistance cassette in community-acquired methicillin-resistant Staphylococcus aureus isolates of diverse genetic backgrounds. *J Infect Dis* 2002;186:1344–47.

2. THE CLOUD BABIES

1. The "Cook-Leichter bill"—written by Constance E. Cook, a Republican assembly-woman from upstate New York, and Franz S. Leichter, a Democrat from Manhattan—revoked prohibitions on abortion that included requiring a woman to find two psychiatrists to first certify that she was a suicide risk. It became effective July 1, 1970. The January 22, 1973, Roe v. Wade decision that made abortion available nationwide was modeled in part on the New York law. For more, see Rosemary Nossiff's *Before Roe* (Philadelphia: Temple University Press, 2001).
2. The description of the Dieringers' experience is based on interviews with Eva (Dieringer) Ferguson on May 10, 2007, and with Eva and her parents, Bev and Ernie Dieringer, on May 20, 2008, and on medical records preserved by the Dieringers.
3. The story of the discovery and development of penicillin is brilliantly told in Eric Lax's *The Mold in Dr. Florey's Coat* (New York: Henry Holt, 2004).
4. Bud R. *Penicillin: Triumph and Tragedy* (UK: Oxford University Press, 2007).
5. Acceptance lecture for the Nobel Prize in Physiology or Medicine. 1945. Accessed January 15, 2008, at http://nobelprize.org/nobel_prizes/medicine/laureates/1945/fleming-lecture.pdf
6. Abraham, 1940.
7. Rammelkamp C, Maxon T. Resistance of Staphylococcus aureus to the action of penicillin. *Proceedings of the Royal Society of Experimental Biology and Medicine* 1942;51:386–89.
8. Kirby WM. Extraction of a highly potent penicillin inactivator from penicillin resistant staphylococci. *Science* 1944;99:452–53.
9. Rountree PM. Infections caused by a particular phage type of Staphylococcus aureus. *Medical Journal of Australia* 1955;42:157–61.
10. Antibiotic Effect Feared on Decline, *New York Times*, July 16, 1953.
11. Laveck GD, Ravenholt RT. Staphylococcal disease; an obstetric, pediatric, and community problem. *Am J Public Health Nations Health* 1956;46:1287–96.
12. Phages are viruses that attack bacteria. In the years immediately after antibiotics were developed, medicine relied on biotyping—identifying which antibiotics bacteria were resistant to—to tell organisms apart. In 1953, a group of British researchers developed phage typing, a quicker and more reliable way of differentiating among bacterial strains. A culture dish of S. aureus would be inoculated with the 23 phages

that were known to react to some staph strains at least. Wherever the bacteria lysed—burst apart and died—was a signal that the strain in the dish was vulnerable to those phages. The staph strain that arose in the 1950s was vulnerable, in a phage-typing test, to phages 80 and 81.

13. Mudd S. Staphylococcic infections in the hospital and community. *American Medical Association Journal* 1958;166:1177–78.

14. Brown JW. Hygiene and education within hospitals to prevent staphylococcic infections. *American Medical Association Journal* 1958;166:1185–91.

15. Burnett WE, Caswell HT, Schreck KM, et al. Program for prevention and eradication of staphylococcic infections. *American Medical Association Journal* 1958;166:1183–84.

16. Godfrey ME, Smith IM. Hospital hazards of staphylococcal sepsis. *American Medical Association Journal* 1958;166:1197–201.

17. Brown, 1958.

18. Blair JE, Carr M. Staphylococci in hospital-acquired infections. *American Medical Association Journal* 1958;166:1192–96.

19. Godfrey, 1958.

20. Details of the Jefferson Davis Hospital epidemic are taken from contemporaneous stories in the *Houston Chronicle* and *Houston Post*, and especially from Mari Nicholson-Preuss's unpublished paper, "A Jefferson Davis Problem: The 1958 Staphylococcus Aureus Epidemic at Houston's Public Charity Hospital," presented at the *Fourth Global Conference Making Sense of Health, Illness and Disease*, at Mansfield College, Oxford, UK, 4–7 July 2005. The paper, part of a forthcoming work, is archived online at http://www.interdisciplinary.net/mso/hid4/Preuss%20paper.pdf.

21. New germ strain takes heavy toll. *New York Times*, March 22, 1958.

22. Nicholson-Preuss, 2005.

23. This section is based on interviews with Dr. Henry Shinefield, May 28, 2008, and Dr. Heinz Eichenwald, January 28, 2009.

24. Eichenwald HF, Shinefield HR. The problem of staphylococcal infection in newborn infants. *J Pediatr* 1960;56:665–74.

25. Davies RR, Noble WC. Dispersal of bacteria on desquamated skin. *Lancet* 1962;2:1295–97.

26. Eichenwald HF, Kotsevalov O, Fasso LA. "The "cloud baby": an example of bacterial-viral interaction." *Am J Dis Child* 1960;100:161–73.

27. The cartoon ran April 21, 1960.

28. Jennings M, Sharp A. Antibacterial activity of the Staphylococcus. *Nature* 1947;159:133–34.

29. Albert H. Diphtheria carriers and their relationship to medical inspection of schools. *Am J Public Health* (NY) 1912;2:794–98.

30. Shinefield HR, Ribble JC, Boris M, Eichenwald HF. Bacterial interference: its effect on nursery-acquired infection with Staphylococcus aureus. I. Preliminary observations on artificial colonzation of newborns. *Am J Dis Child* 1963;105:646–54.

31. Shinefield HR, Sutherland JM, Ribble JC, Eichenwald HF. Bacterial interference: its effect on nursery-acquired infection with Staphylococcus aureus. II. The Ohio epidemic. *Am J Dis Child* 1963;105:655–62.

32. Shinefield HR, Boris M, Ribble JC, Cale EF, Eichenwald HF. Bacterial interference: its effect on nursery-acquired infection with Staphylococcus aureus. III. The Georgia epidemic. *Am J Dis Child* 1963;105:663–73.

33. Boris M, Shinefield HR, Ribble JC, Eichenwald HF, Hauser GH, Caraway CT. Bac-

terial interference: its effect on nursery-acquired infection with Staphylococcus aureus. IV. The Louisiana epidemic. *Am J Dis Child* 1963;105:674–82.

34. Shinefield HR, Ribble JC, Eichenwald HF, Boris M, Sutherland JM. Bacterial interference: its effect on nursery-acquired infection with Staphylococcus aureus. V. An analysis and interpretation. *Am J Dis Child* 1963;105:683–88.

35. Shinefield HR, Ribble JC, Boris M. Bacterial interference between strains of Staphylococcus aureus. *Contrib Microbiol Immunol* 1973;1:541–47.

36. Boris M, Sellers TF, Jr., Eichenwald HF, Ribble JC, Shinefield HR. Bacterial interference; protection of adults against nasal Staphylococcus aureus infection after colonization with a heterologous S. aureus strain. *Am J Dis Child* 1964;108:252–61.

37. Blair EB, Tull AH. Multiple infections among newborns resulting from colonization with Staphylococcus aureus 502A. *Am J Clin Pathol* 1969;52:42–49.

38. Light IJ, Walton RL, Sutherland JM, Shinefield HR, Brackvogel V. Use of bacterial interference to control a staphylococcal nursery outbreak. Deliberate colonization of all infants with the 502A strain of Staphylococcus aureus. *Am J Dis Child* 1967;113:291–300.

39. Drutz DJ, Van Way MH, Schaffner W, Koenig MG. Bacterial interference in the therapy of recurrent staphylococcal infections. Multiple abscesses due to the implantation of the 502A strain of staphylococcus. *N Engl J Med* 1966;275:1161–65.

40. Houck PW, Nelson JD, Kay JL. Fatal septicemia due to Staphylococcus aureus 502A. Report of a case and review of the infectious complications of bacterial interference programs. *Am J Dis Child* 1972;123:45–48.

41. (Advertisement) Celbenin: Effective against all resistant staphylococci. *Quarterly Journal of Medicine,* January 1960.

42. Zierdt CH, Robertson EA, Williams RL, MacLowry JD. Computer analysis of Staphylococcus aureus phage typing data from 1957 to 1975, citing epidemiological trends and natural evolution within the phage typing system. *Appl Environ Microbiol* 1980;39:623–29.

43. The evolution of resistance to penicillin and methicillin is thoroughly discussed in *Revenge of the Microbes: How Bacterial Resistance Is Undermining the Antibiotic Miracle,* by Abigail A. Salyers and Dixie D. Whitt (Washington D.C.: ASM Press, 2005).

44. Jevons, 1961

45. "Celbenin"-resistant staphylococci. *Br Med J* 1961:113–14.

3. THE SILENT ATTACK

1. The Sorensens' experience is reconstructed from interviews with Andrea Sorensen on November 20, 2007; Andrea and Big Jay Sorensen, April 14, 2008; their friend Christine Fusco, April 14, 2008; and Jay's reconstructive surgeon Michael Nagy, MD, January 19, 2009.

2. Hidron AI, Edwards JR, Patel J, et al. NHSN annual update: antimicrobial-resistant pathogens associated with healthcare-associated infections: annual summary of data reported to the National Healthcare Safety Network at the Centers for Disease Control and Prevention, 2006–2007. *Infect Control Hosp Epidemiol* 2008;29:996–1011.

3. Levine, 2006.

4. "Celbenin"-resistant staphylococci. *Br Med J* 1961.

5. Harding JW. Infections due to methicillin-resistant strains of Staphylococcus pyogenes. *J Clin Pathol* 1963;16:268–70.

6. Hallman F. Pathogenic staphylococci from the interior nares: incidence and differentiation. *Proc Soc Exp Biol Med* 1937;36:789-94.

7. Williams RE. Healthy carriage of Staphylococcus aureus: its prevalence and importance. *Bacteriol Rev* 1963;27:56-71.

8. Casewell MW, Hill RL. The carrier state: methicillin-resistant Staphylococcus aureus. *J Antimicrob Chemother* 1986;18 Suppl A:1-12.

9. Williams RE, Jevons MP, Shooter RA, et al. Nasal staphylococci and sepsis in hospital patients. *Br Med J* 1959;2:658-62.

10. Nahmias AJ, Godwin JT, Updyke EL, Hopkins WA. Postsurgical staphylococcic infections. Outbreak traced to an individual carrying phase strains 80/81 and 80/81/52/52A. *JAMA* 1960;174:1269-75.

11. Noble WC, Williams RE, Jevons MP, Shooter RA. Some aspects of nasal carriage of staphylococci. *J Clin Pathol* 1964;17:79-83.

12. Williams, 1963.

13. Stewart GT, Holt RJ. Evolution of natural resistance to the newer penicillins. *Br Med J* 1963;1:308-11.

14. Rountree PM, Beard MA, Loewenthal J, May J, Renwick SB. Staphylococcal sepsis in a new surgical ward. *Br Med J* 1967;1:132-37.

15. Barrett FF, Casey JI, Finland M. Infections and antibiotic use among patients at Boston City Hospital, February, 1967. *N Engl J Med* 1968;278:5-9.

16. Barrett FF, McGehee RF, Jr., Finland M. Methicillin-resistant Staphylococcus aureus at Boston City Hospital. Bacteriologic and epidemiologic observations. *N Engl J Med* 1968;279:441-48.

17. Klimek JJ, Marsik FJ, Bartlett RC, Weir B, Shea P, Quintiliani R. Clinical epidemiologic and bacteriologic observations of an outbreak of methicillin-resistant Staphylococcus aureus at a large community hospital. *Am J Med* 1976;61:340-45.

18. Boyce JM, Landry M, Deetz TR, DuPont HL. Epidemiologic studies of an outbreak of nosocomial methicillin-resistant Staphylococcus aureus infections. *Infect Control* 1981;2:110-16.

19. Peacock JE, Jr., Marsik FJ, Wenzel RP. Methicillin-resistant Staphylococcus aureus: introduction and spread within a hospital. *Ann Intern Med* 1980;93:526-32.

20. Boyce JM. Nosocomial staphylococcal infections. *Ann Intern Med* 1981;95:241-42.

21. Dunkle LM, Naqvi SH, McCallum R, Lofgren JP. Eradication of epidemic methicillin-gentamicin-resistant staphylococcus aureus in an intensive care nursery. *Am J Med* 1981;70:455-58.

22. Saroglou G, Cromer M, Bisno AL. Methicillin-resistant Staphylococcus aureus: interstate spread of nosocomial infections with emergence of gentamicin-methicillin resistant strains. *Infect Control* 1980;1:81-89.

23. Crossley K, Landesman B, Zaske D. An outbreak of infections caused by strains of Staphylococcus aureus resistant to methicillin and aminoglycosides. II. Epidemiologic studies. *J Infect Dis* 1979;139:280-87.

24. Crossley K, Loesch D, Landesman B, Mead K, Chern M, Strate R. An outbreak of infections caused by strains of Staphylococcus aureus resistant to methicillin and aminoglycosides. I. Clinical studies. *J Infect Dis* 1979;139:273-79.

25. Casewell MW. Epidemiology and control of the "modern" methicillin-resistant Staphylococcus aureus. *J Hosp Infect* 1986;7 Suppl A:1-11.

26. Ward TT, Winn RE, Hartstein AI, Sewell DL. Observations relating to an inter-

hospital outbreak of methicillin-resistant Staphylococcus aureus: role of antimicrobial therapy in infection control. *Infect Control* 1981;2:453–59.

27. Thompson RL, Cabezudo I, Wenzel RP. Epidemiology of nosocomial infections caused by methicillin-resistant Staphylococcus aureus. *Ann Intern Med* 1982;97:309–17.

28. Locksley RM, Cohen ML, Quinn TC, et al. Multiply antibiotic-resistant Staphylococcus aureus: introduction, transmission, and evolution of nosocomial infection. *Ann Intern Med* 1982;97:317–24.

29. Panlilio, 1992.

30. Klevens RM, Edwards JR, Tenover FC, McDonald LC, Horan T, Gaynes R. Changes in the epidemiology of methicillin-resistant Staphylococcus aureus in intensive care units in US hospitals, 1992–2003. *Clin Infect Dis* 2006;42:389–91.

31. Elixhauser A, Steiner C. Infections with methicillin-resistant Staphylococcus aureus (MRSA) in US hospitals, 1993–2005. HCUP Statistical Brief #35. Statistical brief. Rockville, MD: Agency for Healthcare Research and Quality; 2007 July.

32. Bancroft EA. Antimicrobial resistance: it's not just for hospitals. *JAMA* 2007;298:1803–4.

4. THE PUZZLE PIECES

1. Borowski J, Kamienska K, Rutecka I. Methicillin-resistant staphylococci. *Br Med J* 1964;1:983.

2. Rhoades ER, Short SG. Methicillin-resistant staphylococci. *Lancet* 1972;1:590.

3. Plorde JJ, Eng SC, Wright LJ, Debesai A. Antibiotic sensitivities of common bacterial pathogens isolated in Addis Ababa; a preliminary report. *Ethiop Med J* 1970;8:107–18.

4. Pal SC, Ray BG. Methicillin-resistant Staphylococci. *J Indian Med Assoc* 1964;42:512–17.

5. Kayser FH. Methicillin-resistant staphylococci 1965–75. *Lancet* 1975;2:650–53.

6. Rosendal K, Jessen O, Bentzon MW, Bulow P. Antibiotic policy and spread of Staphylococcus aureus strains in Danish hospitals, 1969–1974. *Acta Pathol Microbiol Scand* [B] 1977;85:143–52.

7. Pavillard R, Harvey K, Douglas D, et al. Epidemic of hospital-acquired infection due to methicillin-resistant Staphylococcus aureus in major Victorian hospitals. *Med J Aust* 1982;1:451–54.

8. Dominguez MA, de Lencastre H, Linares J, Tomasz A. Spread and maintenance of a dominant methicillin-resistant Staphylococcus aureus (MRSA) clone during an outbreak of MRSA disease in a Spanish hospital. *J Clin Microbiol* 1994;32:2081–87.

9. Gordon RJ, Lowy FD. Pathogenesis of methicillin-resistant Staphylococcus aureus infection. *Clin Infect Dis* 2008;46 Suppl 5:S350–59.

10. Grundmann H, Aires-de-Sousa M, Boyce J, Tiemersma E. Emergence and resurgence of methicillin-resistant Staphylococcus aureus as a public-health threat. *Lancet* 2006;368:874–85.

11. Enright MC, Robinson DA, Randle G, Feil EJ, Grundmann H, Spratt BG. The evolutionary history of methicillin-resistant Staphylococcus aureus (MRSA). *Proc Natl Acad Sci USA* 2002;99:7687–92.

12. Dominguez, 1994.

13. da Silva Coimbra MV, Silva-Carvalho MC, Wisplinghoff H, et al. Clonal spread of methicillin-resistant Staphylococcus aureus in a large geographic area of the United States. *J Hosp Infect* 2003;53:103–10.

14. Crisostomo MI, Westh H, Tomasz A, Chung M, Oliveira DC, de Lencastre H. The evolution of methicillin resistance in Staphylococcus aureus: similarity of genetic backgrounds in historically early methicillin-susceptible and -resistant isolates and contemporary epidemic clones. *Proc Natl Acad Sci USA* 2001;98:9865–70.

15. Barber M. Methicillin-resistant staphylococci. *J Clin Pathol* 1961;14:385–93.

16. Barber M, Waterworth PM. Penicillinase-resistant penicillins and cephalosporins. *Br Med J* 1964;2:344–49.

17. Hartman BJ, Tomasz A. Low-affinity penicillin-binding protein associated with beta-lactam resistance in Staphylococcus aureus. *J Bacteriol* 1984;158:513–16.

18. Murray BE. Problems and dilemmas of antimicrobial resistance. *Pharmacotherapy* 1992;12:86S–93S.

19. Plorde JJ, Sherris JC. Staphylococcal resistance to antibiotics: origin, measurement, and epidemiology. *Ann NY Acad Sci* 1974;236:413–34.

20. Methicillin-resistant staph was so resistant to methicillin, in particular, that the name outlasted the drug. By 2000, methicillin was long out of patent, and its original developers Beecham Research Laboratories no longer existed, having been merged and acquired twice into what became the international pharma company Glaxo-SmithKline. With resistance at such high levels, the drug was not considered valuable, and manufacture of it petered out.

21. Simon Sparrow's story is based on an interview with Everly Macario, December 5, 2007, and an essay she wrote after his death (available at http://www.idsociety.org/Content.aspx?id=5562); emails from her to the author; interviews with the members of the University of Chicago team who treated Simon; and a paper in which he appears without being named: Adem PV, et al., Staphylococcus aureus sepsis and the Waterhouse-Friderichsen syndrome in children. *N Engl J Med*, September 22, 2005;353(12):1245–51.

22. Waterhouse R. A case of suprarenal apoplexy. *Lancet* 1911;1:577–78.

23. Beck WD, Berger-Bachi B, Kayser FH. Additional DNA in methicillin-resistant Staphylococcus aureus and molecular cloning of mec-specific DNA. *J Bacteriol* 1986;165:373–78.

24. Conjugation, transduction, and transformation are thoroughly discussed in Abigail Salyers and Dixie Whitt's *Revenge of the Microbes: How Bacterial Resistance is Undermining the Antibiotic Miracle* (Washington, D.C.: ASM Press, 2005).

25. Katayama Y, Ito T, Hiramatsu K. A new class of genetic element, staphylococcus cassette chromosome mec, encodes methicillin resistance in Staphylococcus aureus. *Antimicrob Agents Chemother* 2000;44:1549–55.

26. Ito T, Katayama Y, Asada K, et al. Structural comparison of three types of staphylococcal cassette chromosome mec integrated in the chromosome in methicillin-resistant Staphylococcus aureus. *Antimicrob Agents Chemother* 2001;45:1323–36.

27. Ma XX, Ito T, Tiensasitorn C, et al. Novel type of staphylococcal cassette chromosome mec identified in community-acquired methicillin-resistant Staphylococcus aureus strains. *Antimicrob Agents Chemother* 2002;46:1147–52.

28. Oliveira DC, Tomasz A, de Lencastre H. Secrets of success of a human pathogen: molecular evolution of pandemic clones of meticillin-resistant Staphylococcus aureus. *Lancet Infect Dis* 2002;2:180–89.

29. Daum, 2002.
30. Purcell K, Fergie JE. Exponential increase in community-acquired methicillin-resistant Staphylococcus aureus infections in South Texas children. *Pediatr Infect Dis J* 2002;21:988–89.
31. Buckingham SC, McDougal LK, Cathey LD, et al. Emergence of community-associated methicillin-resistant Staphylococcus aureus at a Memphis, Tennessee, Children's Hospital. *Pediatr Infect Dis J* 2004;23:619–24.
32. Jungk J, Como-Sabetti K, Stinchfield P, Ackerman P, Harriman K. Epidemiology of methicillin-resistant Staphylococcus aureus at a pediatric healthcare system, 1991–2003. *Pediatr Infect Dis J* 2007;26:339–44.
33. Sattler CA, Mason EO, Jr., Kaplan SL. Prospective comparison of risk factors and demographic and clinical characteristics of community-acquired, methicillin-resistant versus methicillin-susceptible Staphylococcus aureus infection in children. *Pediatr Infect Dis J* 2002;21:910–17.
34. Schultz KD, Fan LL, Pinsky J, et al. The changing face of pleural empyemas in children: epidemiology and management. *Pediatrics* 2004;113:1735–40.
35. Martinez-Aguilar G, Avalos-Mishaan A, Hulten K, Hammerman W, Mason EO, Jr., Kaplan SL. Community-acquired, methicillin-resistant and methicillin-susceptible Staphylococcus aureus musculoskeletal infections in children. *Pediatr Infect Dis J* 2004;23:701–6.
36. Gonzalez BE, Hulten KG, Dishop MK, et al. Pulmonary manifestations in children with invasive community-acquired Staphylococcus aureus infection. *Clin Infect Dis* 2005;41:583–90.
37. Kaplan SL, Hulten KG, Gonzalez BE, et al. Three-year surveillance of community-acquired Staphylococcus aureus infections in children. *Clin Infect Dis* 2005;40:1785–91.
38. Bocchini CE, Hulten KG, Mason EO, Jr., Gonzalez BE, Hammerman WA, Kaplan SL. Panton-Valentine leukocidin genes are associated with enhanced inflammatory response and local disease in acute hematogenous Staphylococcus aureus osteomyelitis in children. *Pediatrics* 2006;117:433–40.
39. Pannaraj PS, Hulten KG, Gonzalez BE, Mason EO, Jr., Kaplan SL. Infective pyomyositis and myositis in children in the era of community-acquired, methicillin-resistant Staphylococcus aureus infection. *Clin Infect Dis* 2006;43:953–60.
40. Gonzalez BE, Teruya J, Mahoney DH, Jr., et al. Venous thrombosis associated with staphylococcal osteomyelitis in children. *Pediatrics* 2006;117:1673–79.
41. Crary SE, Buchanan GR, Drake CE, Journeycake JM. Venous thrombosis and thromboembolism in children with osteomyelitis. *J Pediatr* 2006;149:537–41.
42. Blomquist PH. Methicillin-resistant Staphylococcus aureus infections of the eye and orbit (an American Ophthalmological Society thesis). *Trans Am Ophthalmol Soc* 2006;104:322–45.
43. Arnold SR, Elias D, Buckingham SC, et al. Changing patterns of acute hematogenous osteomyelitis and septic arthritis: emergence of community-associated methicillin-resistant Staphylococcus aureus. *J Pediatr Orthop* 2006;26:703–8.
44. Panton P, Valentine F. Staphylococcal toxin. *Lancet* 1932;1:506–8.
45. Lina G, Piemont Y, Godail-Gamot F, et al. Involvement of Panton-Valentine leukocidin-producing Staphylococcus aureus in primary skin infections and pneumonia. *Clin Infect Dis* 1999;29:1128–32.
46. Gillet Y, Issartel B, Vanhems P, et al. Association between Staphylococcus aureus

strains carrying gene for Panton-Valentine leukocidin and highly lethal necrotising pneumonia in young immunocompetent patients. *Lancet* 2002;359:753–59.

47. Baba T, Takeuchi F, Kuroda M, et al. Genome and virulence determinants of high virulence community-acquired MRSA. *Lancet* 2002;359:1819–27.

48. Vandenesch F, Naimi T, Enright MC, et al. Community-acquired methicillin-resistant Staphylococcus aureus carrying Panton-Valentine leukocidin genes: worldwide emergence. *Emerg Infect Dis* 2003;9:978–84.

49. Robinson DA, Kearns AM, Holmes A, et al. Re-emergence of early pandemic Staphylococcus aureus as a community-acquired methicillin-resistant clone. *Lancet* 2005;365:1256–58.

50. They also found that staph 80/81 had not completely disappeared. When they compared its sequence with modern MRSA strains, they found a high degree of commonality with a community-associated MRSA strain known as the Southwest Pacific clone that had caused serious disease in England and Scotland.

51. Voyich JM, Otto M, Mathema B, et al. Is Panton-Valentine leukocidin the major virulence determinant in community-associated methicillin-resistant Staphylococcus aureus disease? *J Infect Dis* 2006;194:1761–70.

52. Wang R, Braughton KR, Kretschmer D, et al. Identification of novel cytolytic peptides as key virulence determinants for community-associated MRSA. *Nat Med* 2007;13:1510–14.

53. Labandeira-Rey M, Couzon F, Boisset S, et al. Staphylococcus aureus Panton-Valentine leukocidin causes necrotizing pneumonia. *Science* 2007;315:1130–33.

54. Interview with Susan Boyle-Vavra, PhD, Chicago, December 5, 2007.

5. THE BIGGEST THING SINCE AIDS

1. The account of Kepa Askenasy's experience is based on interviews with her on February 29 and July 25, 2008; an interview on July 24, 2008, with a friend who helped take care of her; and a review of her medical records.

2. Perlroth J, Kuo M, Tan J, Bayer AS, Miller LG. Adjunctive use of rifampin for the treatment of Staphylococcus aureus infections: a systematic review of the literature. *Arch Intern Med* 2008;168:805–19.

3. What is PulseNet? 2006. (Accessed March 1, 2009, at http://www.cdc.gov/PULSENET/whatis.htm.)

4. McDougal LK, Steward CD, Killgore GE, Chaitram JM, McAllister SK, Tenover FC. Pulsed-field gel electrophoresis typing of oxacillin-resistant Staphylococcus aureus isolates from the United States: establishing a national database. *J Clin Microbiol* 2003;41:5113–20.

5. Later discoveries would push the number of USA types up to eleven.

6. Centers for Disease Control and Prevention. Methicillin-resistant Staphylococcus aureus skin or soft tissue infections in a state prison—Mississippi, 2000 October 26, 2001; Community-acquired methicillin-resistant Staphylococcus aureus in a rural American Indian community, February 7, 2003.

7. King MD, Humphrey BJ, Wang YF, Kourbatova EV, Ray SM, Blumberg HM. Emergence of community-acquired methicillin-resistant Staphylococcus aureus USA 300 clone as the predominant cause of skin and soft-tissue infections. *Ann Intern Med* 2006;144:309–17.

8. Frazee BW, Lynn J, Charlebois ED, Lambert L, Lowery D, Perdreau-Remington

F. High prevalence of methicillin-resistant Staphylococcus aureus in emergency department skin and soft tissue infections. *Ann Emerg Med* 2005;45:311–20.

9. Moran GJ, Amii RN, Abrahamian FM, Talan DA. Methicillin-resistant Staphylococcus aureus in community-acquired skin infections. *Emerg Infect Dis* 2005;11:928–30.

10. Johnson JK, Khoie T, Shurland S, Kreisel K, Stine OC, Roghmann MC. Skin and soft tissue infections caused by methicillin-resistant Staphylococcus aureus USA300 clone. *Emerg Infect Dis* 2007;13:1195–1200.

11. Moran GJ, Krishnadasan A, Gorwitz RJ, et al. Methicillin-resistant S. aureus infections among patients in the emergency department. *N Engl J Med* 2006;355:666–74.

12. Miller LG, Perdreau-Remington F, Rieg G, et al. Necrotizing fasciitis caused by community-associated methicillin-resistant Staphylococcus aureus in Los Angeles. *N Engl J Med* 2005;352:1445–53.

13. In addition to the articles cited, the account of University of California–San Francisco's work on MRSA is drawn from interviews with Dr. Henry "Chip" Chambers (July 23 and 25, 2008), Francoise Perdreau-Remington, PhD (July 24, 2008), Binh An Diep, PhD (November 24, 2008), and Dr. Adam Hersh (December 15, 2008).

14. Pan ES, Diep BA, Carleton HA, et al. Increasing prevalence of methicillin-resistant Staphylococcus aureus infection in California jails. *Clin Infect Dis* 2003;37:1384–88.

15. Diep BA, Sensabaugh GF, Somboona NS, Carleton HA, Perdreau-Remington F. Widespread skin and soft-tissue infections due to two methicillin-resistant Staphylococcus aureus strains harboring the genes for Panton-Valentine leukocidin. *J Clin Microbiol* 2004;42:2080–84.

16. Diep BA, Carleton HA, Chang RF, Sensabaugh GF, Perdreau-Remington F. Roles of 34 virulence genes in the evolution of hospital- and community-associated strains of methicillin-resistant Staphylococcus aureus. *J Infect Dis* 2006;193:1495–503.

17. Liu C, Graber CJ, Karr M, et al. A population-based study of the incidence and molecular epidemiology of methicillin-resistant Staphylococcus aureus disease in San Francisco, 2004–2005. *Clin Infect Dis* 2008;46:1637–46.

18. Fridkin SK, Hageman JC, Morrison M, et al. Methicillin-resistant Staphylococcus aureus disease in three communities. *N Engl J Med* 2005;352:1436–44.

19. Stryjewski ME, Chambers HF. Skin and soft-tissue infections caused by community-acquired methicillin-resistant Staphylococcus aureus. *Clin Infect Dis* 2008;46 Suppl 5:S368–77.

20. Charlebois ED, Bangsberg DR, Moss NJ, et al. Population-based community prevalence of methicillin-resistant Staphylococcus aureus in the urban poor of San Francisco. *Clin Infect Dis* 2002;34:425–33.

21. Sifri CD, Park J, Helm GA, Stemper ME, Shukla SK. Fatal brain abscess due to community-associated methicillin-resistant Staphylococcus aureus strain USA300. *Clin Infect Dis* 2007;45:e113–17.

22. Lewis JS, 2nd, Jorgensen JH. Inducible clindamycin resistance in Staphylococci: should clinicians and microbiologists be concerned? *Clin Infect Dis* 2005;40:280–85.

23. Diep BA, Gill SR, Chang RF, et al. Complete genome sequence of USA300, an epidemic clone of community-acquired meticillin-resistant Staphylococcus aureus. *Lancet* 2006;367:731–39.

24. Tenover FC, McDougal LK, Goering RV, et al. Characterization of a strain of community-associated methicillin-resistant Staphylococcus aureus widely disseminated in the United States. *J Clin Microbiol* 2006;44:108–18.

25. Diep BA, Stone GG, Basuino L, et al. The arginine catabolic mobile element and

staphylococcal chromosomal cassette mec linkage: convergence of virulence and resistance in the USA300 clone of methicillin-resistant Staphylococcus aureus. *J Infect Dis* 2008;197:1523–30.

26. Kravitz GR, Dries DJ, Peterson ML, Schlievert PM. Purpura fulminans due to Staphylococcus aureus. *Clin Infect Dis* 2005;40:941–47.

27. Interview with Patrick Schlievert, PhD, February 26, 2008.

28. Klevens RM, Morrison MA, Nadle J, et al. Invasive methicillin-resistant Staphylococcus aureus infections in the United States. *JAMA* 2007;298:1763–71.

29. Carleton HA, Diep BA, Charlebois ED, Sensabaugh GF, Perdreau-Remington F. Community-adapted methicillin-resistant Staphylococcus aureus (MRSA): population dynamics of an expanding community reservoir of MRSA. *J Infect Dis* 2004;190:1730–38.

30. Liu, 2008.

31. Kennedy AD, Otto M, Braughton KR, et al. Epidemic community-associated methicillin-resistant Staphylococcus aureus: recent clonal expansion and diversification. *Proc Natl Acad Sci USA* 2008;105:1327–32.

32. McDougal LK, Fosheim G, Patel J. Emergence of resistance among USA300 MRSA isolates causing invasive disease in the U.S. C1–166. *48th Annual Interscience Conference on Antimicrobial Agents and Chemotherapy and the Infectious Diseases Society of America 46th Annual Meeting,* October 25–28, 2008, Washington, D.C.

6. THE FLOOD RISING

1. Kennedy L, Gill J, Schultz M, Irmler M, Gordin F. Inside-Out: The Changing Epidemiology of MRSA.276. *45th Annual Meeting of the Infectious Diseases Society of America,* October 4–7, 2007, San Diego, CA.

2. Bebinger D, Ellison R, III, Chowdry R, Erlichman R. Epidemiology of Community Associated MRSA Infections at a Massachusetts Tertiary Care Center. *17th Annual Scientific Meeting of the Society for Healthcare Epidemiology of America,* April 14–17, 2007, Baltimore, MD.

3. Kainer M, Omohundro E. Increase in Burden of Skin and Soft Tissue Infections (SSTI) as Measured by Emergency Department (ED) Visits, 23 Hour Observations and Ambulatory Surgical Center Visits, Tennessee (TN) 2000–2004, *17th Annual Scientific Meeting of the Society for Healthcare Epidemiology of America,* April 14–17, 2007, Baltimore, MD.

4. Chambers, July 23, 2008.

5. The International Standard Classification of Diseases and Related Health Problems, usually known by the abbreviated acronym ICD, is online at http://www.who.int/classifications/icd/en/.

6. Hersh AL, Chambers HF, Maselli JH, Gonzales R. National trends in ambulatory visits and antibiotic prescribing for skin and soft-tissue infections. *Arch Intern Med* 2008;168:1585–91.

7. Interview with Adam Hersh, December 15, 2008.

8. Strategies for Clinical Management of MRSA in the Community: Summary of an Experts' Meeting Convened by the Centers for Disease Control and Prevention, Centers for Disease Control and Prevention, 2006. (Accessed at http://www.cdc.gov/ncidod/dhqp/ar_mrsa_ca.html.)

9. The account of Steve and Sue McNees's illness is based on a joint interview with

them on January 2, 2008; emails from them; and emails from Steve's sister Pat McNees.

10. Williams, 1963.

11. Kuehnert, 2006.

12. Gorwitz RJ, Kruszon-Moran D, McAllister SK, et al. Changes in the prevalence of nasal colonization with Staphylococcus aureus in the United States, 2001–2004. *J Infect Dis* 2008;197:1226–34.

13. Creech CB, II, Kernodle DS, Alsentzer A, Wilson C, Edwards KM. Increasing rates of nasal carriage of methicillin-resistant Staphylococcus aureus in healthy children. *Pediatr Infect Dis J* 2005;24:617–21.

14. Ellis MW, Hospenthal DR, Dooley DP, Gray PJ, Murray CK. Natural history of community-acquired methicillin-resistant Staphylococcus aureus colonization and infection in soldiers. *Clin Infect Dis* 2004;39:971–79.

15. Patel G, Radbill B, Jenkins S, Calfee D. Methicillin resistant Staphylococcus aureus carriage in ambulatory hemodialysis patients. K-538. *48th Annual Interscience Conference on Antimicrobial Agents and Chemotherapy and the Infectious Diseases Society of America 46th Annual Meeting,* October 25–28, 2008, Washington, D.C.

16. Chen KT, Campbell H, Borrell LN, Huard RC, Saiman L, Della-Latta P. Predictors and outcomes for pregnant women with vaginal-rectal carriage of community-associated methicillin-resistant Staphylococcus aureus. *Am J Perinatol* 2007;24:235–40.

17. Yang ES, Tan J, Rieg G, Miller LG. Body site colonization prevalence in patients with community-associated methicillin-resistant Staphylococcus aureus infections. *45th Annual Meeting of the Infectious Diseases Society of America,* October 4–7, 2007, San Diego, CA.

18. Mertz D, Frei R, Jaussi B, et al. Throat swabs are necessary to reliably detect carriers of Staphylococcus aureus. *Clin Infect Dis* 2007;45:475–77.

19. Creech CB, Beekmann SE, Chen Y, Polgreen PM. Variability among pediatric infectious diseases specialists in the treatment and prevention of methicillin-resistant Staphylococcus aureus skin and soft tissue infections. *Pediatr Infect Dis J* 2008;27:270–72.

20. Henderson RJ, Williams RE. Nasal disinfection in prevention of postoperative staphylococcal infection of wounds. *Br Med J* 1961;2:330–33.

21. Ammerlaan HS, Kluytmans JA, Wertheim HF, Nouwen JL, Bonten MJ. Eradication of methicillin-resistant Staphylococcus aureus carriage: a systematic review. *Clin Infect Dis* 2009;48:922–30.

22. CDC "Strategies for Clinical Management" brief, 2008.

23. The account of the Logan family's experience with MRSA is based on interviews with Mollie Logan on July 10, August 15, and November 2, 2006, and on an interview with Darcy Jones, PA of Omaha Infectious Disease Associates, who treated Mollie, on July 24, 2006. Portions of the Logans' story initially appeared in *Self* magazine, February 2007.

24. Jones TF, Creech CB, Erwin P, Baird SG, Woron AM, Schaffner W. Family outbreaks of invasive community-associated methicillin-resistant Staphylococcus aureus infection. *Clin Infect Dis* 2006;42:e76–78.

25. Based on an interview with Carlo Vitale, DVM, in San Francisco on July 24, 2008; and on his paper "Methicillin-resistant Staphylococcus aureus in cat and owner," *Emerg Infect Dis* 2006;1998–2000.

26. Scott GM, Thomson R, Malone-Lee J, Ridgway GL. Cross-infection between ani-

mals and man: possible feline transmission of Staphylococcus aureus infection in humans? *J Hosp Infect* 1988;12:29–34.

27. Vitale CB, Gross TL, Weese JS. Methicillin-resistant Staphylococcus aureus in cat and owner. *Emerg Infect Dis* 2006;12:1998–2000.

28. Cefai C, Ashurst S, Owens C. Human carriage of methicillin-resistant Staphylococcus aureus linked with pet dog. *Lancet* 1994;344:539–40.

29. Manian FA. Asymptomatic nasal carriage of mupirocin-resistant, methicillin-resistant Staphylococcus aureus (MRSA) in a pet dog associated with MRSA infection in household contacts. *Clin Infect Dis* 2003;36:e26–28.

30. National Pet Owners Survey 2007–2008. 2008. (Accessed January 16, 2009, at http://americanpetproducts.org/press_industrytrends.asp.)

31. Baptiste KE, Williams K, Williams NJ, et al. Methicillin-resistant staphylococci in companion animals. *Emerg Infect Dis* 2005;11:1942–44.

32. Moodley A, Stegger M, Bagcigil AF, et al. Spa typing of methicillin-resistant Staphylococcus aureus isolated from domestic animals and veterinary staff in the UK and Ireland. *J Antimicrob Chemother* 2006;58:1118–23.

33. Loeffler A, Boag AK, Sung J, et al. Prevalence of methicillin-resistant Staphylococcus aureus among staff and pets in a small animal referral hospital in the UK. *J Antimicrob Chemother* 2005;56:692–97.

34. Weese JS, Caldwell F, Willey BM, et al. An outbreak of methicillin-resistant Staphylococcus aureus skin infections resulting from horse to human transmission in a veterinary hospital. *Vet Microbiol* 2006;114:160–64.

35. Hanselman BA, Kruth SA, Rousseau J, et al. Methicillin-resistant Staphylococcus aureus colonization in veterinary personnel. *Emerg Infect Dis* 2006;12:1933–38.

36. Croft DR, Sotir MJ, Williams CJ, et al. Occupational risks during a monkeypox outbreak, Wisconsin, 2003. *Emerg Infect Dis* 2007;13:1150–57.

37. van Kolfschooten F. Dutch veterinarian becomes first victim of avian influenza. *Lancet* 2003;361:1444.

38. Wright JG, Jung S, Holman RC, Marano NN, McQuiston JH. Infection control practices and zoonotic disease risks among veterinarians in the United States. *J Am Vet Med Assoc* 2008;232:1863–72.

39. Steele J. Wild Animal Park forced to euthanize elephant. *San Diego Union-Tribune*, 2008 February 6;Sect. 3.

40. Steinberg J. US agency's report could help complete elephant import deal. *San Diego Union-Tribune*, 2003 June 7;Sect. 5–7.

41. African Elephants at the Wild Animal Park. 2009. (Accessed January 13, 2009, at http://www.sandiegozoo.org/news/gallery_baby-elephant.html.)

42. This section is based on interviews with Dr. Don Janssen and Dr. Nadine Lamberski of the San Diego Zoo's Wild Animal Park on December 12, 2008, and with Jeff Andrews, animal care manager for the elephant herd, on January 16, 2009.

43. Clubb R, Rowcliffe M, Lee P, Mar KU, Moss C, Mason GJ. Compromised survivorship in zoo elephants. *Science* 2008;322:1649.

44. Sugerman DE, Janssen D, Lamberski N, et al. Community-associated MRSA outbreak involving an elephant calf—San Diego, 2008. *48th Annual Interscience Conference on Antimicrobial Agents and Chemotherapy and the Infectious Diseases Society of America 46th Annual Meeting*, October 25–28, 2008, Washington, D.C.

45. Centers for Disease Control and Prevention. Methicillin-resistant Staphylococcus

aureus skin infections from an elephant calf—San Diego, CA, 2008. *MMWR, Morb Mortal Wkly Rep* 2009;58:194–98.

7. GONE WITHOUT WARNING

1. The description of Bryce Smith's experience is based on interviews with Scott and Katie Smith on October 7, 2007; and on interviews with Bryce's treatment team, including Dr. John Bradley (July 11 and October 8, 2007) and Dr. Bradley Peterson (October 9, 2007).

2. Mandell LA, Bartlett JG, Dowell SF, File TM, Jr., Musher DM, Whitney C. Update of practice guidelines for the management of community-acquired pneumonia in immunocompetent adults. *Clin Infect Dis* 2003;37:1405–33.

3. Dufour P, Gillet Y, Bes M, et al. Community-acquired methicillin-resistant Staphylococcus aureus infections in France: emergence of a single clone that produces Panton-Valentine leukocidin. *Clin Infect Dis* 2002;35:819–24.

4. Klein JL, Petrovic Z, Treacher D, Edgeworth J. Severe community-acquired pneumonia caused by Panton-Valentine leukocidin-positive Staphylococcus aureus: first reported case in the United Kingdom. *Intensive Care Med* 2003;29:1399.

5. Peleg AY, Munckhof WJ. Fatal necrotising pneumonia due to community-acquired methicillin-resistant Staphylococcus aureus (MRSA). *Med J Aust* 2004;181:228–29.

6. Francis JS, Doherty MC, Lopatin U, et al. Severe community-onset pneumonia in healthy adults caused by methicillin-resistant Staphylococcus aureus carrying the Panton-Valentine leukocidin genes. *Clin Infect Dis* 2005;40:100–107.

7. Frazee BW, Salz TO, Lambert L, Perdreau-Remington F. Fatal community-associated methicillin-resistant Staphylococcus aureus pneumonia in an immunocompetent young adult. *Ann Emerg Med* 2005;46:401–4.

8. Kilbourne ED. *Influenza* (New York: Plenum, 1987).

9. Grist NR. Pandemic influenza 1918. *Br Med J* 1979;2:1632–33.

10. Louria DB, Blumenfeld HL, Ellis JT, Kilbourne ED, Rogers DE. Studies on influenza in the pandemic of 1957–1958. II. Pulmonary complications of influenza. *J Clin Invest* 1959;38:213–65.

11. Schwarzmann SW, Adler JL, Sullivan RJ, Jr., Marine WM. Bacterial pneumonia during the Hong Kong influenza epidemic of 1968–1969. *Arch Intern Med* 1971;127:1037–41.

12. Bacterial coinfections in lung tissue specimens from fatal cases of 2009 pandemic influenza A (H1N1)—United States, May–August 2009. *MMWR, Morb Mortal Wkly Rep* 2009;58:1071–74.

13. Taubenberger JK, Reid AH, Krafft AE, Bijwaard KE, Fanning TG. Initial genetic characterization of the 1918 "Spanish" influenza virus. *Science* 1997;275:1793–96.

14. Morens DM, Taubenberger JK, Fauci AS. Predominant role of bacterial pneumonia as a cause of death in pandemic influenza: implications for pandemic influenza preparedness. *J Infect Dis* 2008;198:962–70.

15. McCullers JA. Insights into the interaction between influenza virus and pneumococcus. *Clin Microbiol Rev* 2006;19:571–82.

16. Auge K. Three more Colorado kids die from flu. *Denver Post* 2003 November 26; Sect. 1.

17. Associated Press. CDC to monitor children's flu complications. 2003 December 9.

18. Centers for Disease Control and Prevention. Update: Influenza activity—United States, 2003–04 season. *MMWR, Morb Mortal Wkly Rep* 2003;52:1197–202.

19. Bhat N, Wright JG, Broder KR, et al. Influenza-associated deaths among children in the United States, 2003–2004. *N Engl J Med* 2005;353:2559–67.

20. Hageman JC, Uyeki TM, Francis JS, et al. Severe community-acquired pneumonia due to Staphylococcus aureus, 2003–04 influenza season. *Emerg Infect Dis* 2006;12:894–99.

21. Interview with Jeffrey Hageman, March 28, 2008.

22. Whitney CG, Farley MM, Hadler J, et al. Decline in invasive pneumococcal disease after the introduction of protein-polysaccharide conjugate vaccine. *N Engl J Med* 2003;348:1737–46.

23. Kallen AJ, Brunkard J, Moore Z, et al. Staphylococcus aureus community-acquired pneumonia during the 2006 to 2007 influenza season. *Ann Emerg Med* 2009;53:358–65.

24. Centers for Disease Control and Prevention. Severe methicillin-resistant Staphylococcus aureus community-acquired pneumonia associated with influenza—Louisiana and Georgia, December 2006–January 2007. *MMWR, Morb Mortal Wkly Rep* 2007;56:325–29.

25. Goslings WR, Mulder J, Djajadiningrat J, Masurel N. Staphylococcal pneumonia in influenza in relation to antecedent staphylococcal skin infection. *Lancet* 1959;2:428–30.

26. Kallen, 2008.

27. Fiore AE, Shay DK, Haber P, et al. Prevention and control of influenza. Recommendations of the Advisory Committee on Immunization Practices (ACIP), 2007. *MMWR Recomm Rep* 2007;56:1–54.

28. Centers for Disease Control and Prevention. Influenza vaccination coverage among children aged 6–59 months—Eight immunization information system sentinel sites, United States, 2007–08 influenza season. *MMWR, Morb Mortal Wkly Rep* 2008;57:1043–46.

29. Carlos Don's experience is based on interviews with Amber Don (July 6 and October 8, 2007) and with his physicians Dr. John Bradley (July 11 and October 8, 2007 and March 12, 2009), and Dr. Bradley Peterson (October 9, 2007); and on an account later written by Amber Don.

8. THE FOE ON THE FIELD

1. The account of Nick Johnson's experience is based on interviews with David Edell (October 6 and November 10, 2008); Janet Johnson (October 10 and November 7, 2008); and Nick Johnson (November 7, 2008).

2. In addition to interviews with Nick Johnson, Janet Johnson, and Dave Edell, the account of Nick Johnson's treatment is based on a narrative of Nick's case written up as a teaching tool by physicians Marianna Sockrider, MD, DrPH, and Morven Edwards, MD, at Baylor College of Medicine ("Integrated Problem Solving, Spring 2004: Nick Johnson.")

3. Stanley D. Stubborn staph infections spread to UT athletes, others in Texas. *Austin* (TX) *American-Statesman* 2002 December 20;Sect. 1.

4. Allen M. Resistant bacteria spur concerns. *San Antonio* (TX) *Express-News* 2002 December 16;Sect. 1.

5. Lindenmayer JM, Schoenfeld S, O'Grady R, Carney JK. Methicillin-resistant Staphylococcus aureus in a high school wrestling team and the surrounding community. *Arch Intern Med* 1998;158:895–99.

6. Interview with Joann Lindenmayer, DVM, MPH, December 8, 2008.

7. Mihoces G. Skin infection bacteria requires vigilance. *USA Today* 2003 October 15;Sect. 12.

8. Epstein V. Texas football succumbs to virulent staph infection from turf. *Bloomberg News* 2007 December 21.

9. Trubow A. In dollars, coaches beat teachers. *Austin American-Statesman* 2006 August 27;Sect. 1.

10. Callan S, Thomas J. Pay and performance: an examination of Texas high school football coaches. *The Sport Journal* 2008;11.

11. Jacob M. High school football coaches cashing in. *Dallas Morning News* 2006.

12. Knight P. No pressure. It's just Texas high school football. *Houston Press* 2008 October 21.

13. Barr B, Felkner M, Diamond P. High school athletic departments as sentinel surveillance sites for community-associated methicillin-resistant staphylococcal infections. *Texas Medicine* 2006;102:56–61.

14. Begier EM, Frenette K, Barrett NL, et al. A high-morbidity outbreak of methicillin-resistant Staphylococcus aureus among players on a college football team, facilitated by cosmetic body shaving and turf burns. *Clin Infect Dis* 2004;39:1446–53.

15. Kazakova SV, Hageman JC, Matava M, et al. A clone of methicillin-resistant Staphylococcus aureus among professional football players. *N Engl J Med* 2005;352:468–75.

16. Centers for Disease Control and Prevention. Methicillin-resistant staphylococcus aureus infections among competitive sports participants—Colorado, Indiana, Pennsylvania, and Los Angeles County, 2000–2003. *MMWR, Morb Mortal Wkly Rep* 2003;52:793–95.

17. Nguyen DM, Mascola L, Brancoft E. Recurring methicillin-resistant Staphylococcus aureus infections in a football team. *Emerg Infect Dis* 2005;11:526–32.

18. Romano R, Lu D, Holtom P. Outbreak of community-acquired methicillin-resistant Staphylococcus aureus skin infections among a collegiate football team. *J Athl Train* 2006;41:141–45.

19. MRSA has sidelined careers, even caused death. ESPN.com, 2007. (Accessed December 10, 2008, at http://sports.espn.go.com/espn/news/story?id=2800948.)

20. Jarrett K. UNCA center Kenny George loses part of foot. *Asheville Citizen-Times* 2008 October 17.

21. Withers T. Growing threat: Staph infections rising among athletes. Associated Press 2006 November 22.

22. Arangure Jr. J. Sosa wants to end career as an Oriole. *Washington Post* 2005 June 10;Sect. 11.

23. Gillies R. Jays disinfect clubhouse after second player has staph infection. Associated Press 2006 July 16.

24. Maske M, LaCanfora J. A frightening off-field foe. *Washington Post* 2006 January 27;Sect. 1.

25. NFL stars no match for bacteria. CNN.com, 2008. (Accessed December 16, 2008, at http://edition.cnn.com/2008/HEALTH/conditions/10/28/staph.football/index.html.)

26. Ricky Lannetti's story. Infectious Diseases Society of America, 2007. (Accessed September 11, 2007, at http://www.idsociety.org/Content.aspx?id=5626.)

27. Bryant H. Blitzing microbial infections; Redskins take multiple precautions against dangerous MRSA. *Washington Post* 2006 August 3;Sect. 1.
28. This section is based on interviews with Tom Keating on March 10 and May 12, 2008, and on materials he developed for Project CLEAN, www.project-clean.com. He shared his bathroom rounds on the condition that school names not be disclosed.
29. The CDC's recommendations for dealing with MRSA in athletic settings are posted at http://www.cdc.gov/ncidod/dhqp/ar_MRSA_AthletesFAQ.html
30. Cutter H. Mandatory showers strip students of rights, ACLU says. Associated Press 1994 November 12.
31. Dizon K. Raising the bar. *Seattle Post-Intelligencer* 2001 May 7.
32. Dotinga R. Where have all the showers gone? *Christian Science Monitor* 2004 February 3.
33. Herzog K. The shower police. *Milwaukee Sentinel* 1995 January 13.
34. Owen W. Shower together at school? No way, dude. *The Oregonian* 2007 December 2;Sect. 1.

9. PRISON, THE PERFECT INCUBATOR

1. The sections on Kevin Keller's experience and the Bucks County Correctional Facility are based on interviews with Keller (April 10, 2009) and his attorneys Anita Alberts (January 28, 2009) and Martha Sperling (February 24, 2009); and on the various filings for Inmates of the Bucks County Correctional Facility v. County of Bucks (2004 WL 2958427, (E.D.Pa., 2004)), Keller v. County of Bucks (2005 WL 675.831 (E.D. Pa. 2005)), and Report of the Bucks County Task Force on Incarcerated Women and the Mentally Ill (2001).
2. In addition to the articles cited, the sections on the emergence of MRSA in jails and prisons are based on interviews with Dr. Elizabeth Bancroft, Los Angeles County Department of Public Health (July 24, 2006, and November 27, 2007); Dr. John Clarke, former medical director, Los Angeles County Sheriff's Department (March 23, 2009); attorney Gregory Belzley (January 2, 2009); and Timothy Faloon, Rhonda Moore, RN, and Dr. Luis Rivera of the Anne Arundel County Department of Detention Facilities (October 29, 2008).
3. Centers for Disease Control and Prevention. Methicillin-resistant Staphylococcus aureus skin or soft tissue infections in a state prison—Mississippi, October 26, 2001.
4. Centers for Disease Control and Prevention. Methicillin-resistant Staphylococcus aureus infections in correctional facilities—Georgia, California, and Texas, 2001–2003. *MMWR, Morb Mortal Wkly Rep* 2003;52:992–96.
5. Centers for Disease Control and Prevention, February 7, 2003 and October 17, 2003.
6. Centers for Disease Control and Prevention, October 17, 2003.
7. Bick JA. Infection control in jails and prisons. *Clin Infect Dis* 2007;45:1047–55.
8. Pan, Increasing prevalence . . . Staphylococcus . . . in California jails, 2003.
9. Brunkard J, Scott CM, Haydel D, McAllister SK, Ratard RC. High prevalence of methicillin-resistant Staphylococcus aureus nasal colonization among inmates in a New Orleans parish jail, 2007–2008. L-1494 *48th Annual Interscience Conference on Antimicrobial Agents and Chemotherapy and the Infectious Diseases Society of America 46th Annual Meeting*, Washington, D.C., October 25–28, 2008.
10. Lowy FD, Aiello AE, Bhat M, et al. Staphylococcus aureus colonization and infection in New York State prisons. *J Infect Dis* 2007;196:911–18.

11. Wright MO, Furuno JP, Venezia RA, et al. Methicillin-resistant Staphylococcus aureus infection and colonization among hospitalized prisoners. *Infect Control Hosp Epidemiol* 2007;28:877–79.

12. Felkner M. Methicillin-resistant Staphylococcus aureus nasal carriage rate in Texas county jail inmates. *Journal of Correctional Health Care* 2007;13:289–95.

13. Prison Inmates at Midyear 2008—Statistical Tables. Office of Justice Programs, Department of Justice, 2009. (Accessed April 15, 2009, at http://www.ojp.usdoj.gov/bjs/abstract/pim08st.htm.)

14. Jail Inmates at Midyear 2008—Statistical Tables. Office of Justice Programs, Department of Justice, 2009. (Accessed April 15, 2009, at http://www.ojp.usdoj.gov/bjs/abstract/jim08st.htm.)

15. This section on James Bell is based on an exchange of letters with him between February and April 2009, on an interview with Nate Wenstrup of the Ohio Justice and Policy Center on December 26, 2008, and on news coverage of Bell's case.

16. Riepenhoff J. Five prisoners face charges from Lucasville riot. *Columbus Dispatch* 1994 February 4;Sect. 1.

17. Stipulation for Injunctive Relief, Fussell v. Wilkinson, No. C-1-03-704 (S.D. Ohio, November 22, 2005).

18. Ludlow R. Danger of staph outbreak realized too late. *Columbus Dispatch* 2003 May 20;Sect. 1.

19. Mueller E. Ex-inmate with skin lesions dies; Death may be second one resulting from infection at Pickaway County prison. *Columbus Dispatch* 2003 May 23;Sect. 1.

20. Fussell v. Wilkinson, US District Court, Southern District of Ohio, Western Division, Case No. C-1-03-704. ODRC Medical Services Final Report of Findings.

21. Confronting Confinement. Commission on Safety and Abuse in America's Prisons, 2006. (Accessed February 2, 2009, at http://www.prisoncommission.org.)

22. Hartley DM, Furuno JP, Wright MO, Smith DL, Perencevich EN. The role of institutional epidemiologic weight in guiding infection surveillance and control in community and hospital populations. *Infect Control Hosp Epidemiol* 2006;27:170–74.

23. Hota B, Ellenbogen C, Hayden MK, Aroutcheva A, Rice TW, Weinstein RA. Community-associated methicillin-resistant Staphylococcus aureus skin and soft tissue infections at a public hospital: do public housing and incarceration amplify transmission? *Arch Intern Med* 2007;167:1026–33.

24. Interview with Gregory Belzley, Esq., January 2, 2009.

25. The description of the Folsom State Prison guards' experience is based on interviews with Chris Pearson (March 25, 2009), Gary Benson (April 4, 2009), and David White (April 2, 2009); with Charles Alexander, Jr., state executive vice president of the California Correctional Peace Officers Association (February 9, 2009); with the guards' attorney, Natalie Leonard of Carroll, Burdick & McDonough LLP, San Francisco; on written opinions by the State of California Division of Occupational Safety and Health; and on various filings in the guards' suit in the Superior Court of California, County of San Francisco: California Correctional Peace Officers et al. vs. Matthew C. Kramer (case CGC-07-463194).

26. The receivership of the prison health system was created by the case Plata v. Schwarzenegger et al. A complete archive of documents related to California prison health is maintained by the state government, as a result of the case, at: http://www.cprinc.org/plata.aspx

27. Chuck Helton, Vice President, Adult Division, California Correctional Peace Offi-

cers Association, to Suzan Hubbard, California Department of Corrections and Rehabilitation, Letter "Re CCPOA MS #19556/CDCR MRSA Recognition," April 24, 2008.

10. INTO THE FOOD CHAIN

1. Interview with Tara C. Smith, PhD, Gregory Gray, MD, and MacDonald Farnham, DVM, August 28, 2008.
2. Harper AL, Male MJ, Moritz-Korolev E, Diekema DJ, Herwaldt LA, Smith TC. Prevalence of methicillin-resistant Staphylococcus aureus (MRSA) in swine from the midwestern United States. Poster 0983. American Society for Microbiology 108th General Meeting, June 1–5, 2008, Boston.
3. Voss A, Loeffen F, Bakker J, Klaassen C, Wulf M. Methicillin-resistant Staphylococcus aureus in pig farming. Emerg Infect Dis 2005;11:1965–66.
4. Vandenbroucke-Grauls CM. Methicillin-resistant Staphylococcus aureus control in hospitals: the Dutch experience. Infect Control Hosp Epidemiol 1996;17:512–13.
5. MRSA Ziekenhuis [Guidelines for MRSA in Hospitals]. 2007. (Accessed December 1, 2008, at http://www.wip.nl/UK/free_content/Richtlijnen/MRSA%20hospital.pdf.)
6. Wertheim HF, Vos MC, Boelens HA, et al. Low prevalence of methicillin-resistant Staphylococcus aureus (MRSA) at hospital admission in the Netherlands: the value of search and destroy and restrictive antibiotic use. J Hosp Infect 2004;56:321–25.
7. Salgado CD, Farr BM, Calfee DP. Community-acquired methicillin-resistant Staphylococcus aureus: a meta-analysis of prevalence and risk factors. Clin Infect Dis 2003;36:131–39.
8. Wertheim, 2004.
9. Interview with Professor Andreas Voss, January 22, 2009.
10. SWAB guidelines for the treatment of MRSA carriage. 2007. (Accessed January 3, 2009, at http://www.swab.nl/swab/swabcms.nsf/(WebFiles)/719CF19988D366EFC1 2574F3004E3F14/$FILE/MRSA%20Richtlijn%20engelse%20versie%20311008.pdf.)
11. Crossley, Staphylococci in Human Disease, pp. 406–7.
12. Lee JH. Methicillin (Oxacillin)-resistant Staphylococcus aureus strains isolated from major food animals and their potential transmission to humans. Appl Environ Microbial 2003;69:6489–94.
13. Bens CC, Voss A, Klaassen CH. Presence of a novel DNA methylation enzyme in methicillin-resistant Staphylococcus aureus isolates associated with pig farming leads to uninterpretable results in standard pulsed-field gel electrophoresis analysis. J Clin Microbiol 2006;44:1875–76.
14. Huijsdens XW, van Dijke BJ, Spalburg E, et al. Community-acquired MRSA and pig-farming. Ann Clin Microbiol Antimicrob 2006;5:26.
15. Ekkelenkamp MB, Sekkat M, Carpaij N, Troelstra A, Bonten MJ. [Endocarditis due to meticillin-resistant Staphylococcus aureus originating from pigs]. Ned Tijdschr Geneeskd 2006;150:2442–47.
16. de Neeling AJ, van den Broek MJ, Spalburg EC, et al. High prevalence of methicillin-resistant Staphylococcus aureus in pigs. Vet Microbiol 2007;122:366–72.
17. Wulf M, van Nes A, Eikelenboom-Boskamp A, et al. Methicillin-resistant Staphylococcus aureus in veterinary doctors and students, the Netherlands. Emerg Infect Dis 2006;12:1939–41.
18. Huijsdens X, Bosch T, Haenen A, et al. Molecular epidemiology of PFGE non-

typeable methicillin-resistant Staphylococcus aureus in the Netherlands. P1407. *18th European Congress of Clinical Microbiology and Infectious Diseases*, April 19, 2008, Barcelona.

19. First outbreak of methicillin-resistant Staphylococcus aureus ST398 in a Dutch hospital, June 2007. 2008. (Accessed October 22, 2008, at http://www.eurosurveillance.org/ViewArticle.aspx?ArticleId=8051.)

20. Wulf MW, Tiemersma E, Kluytmans J, et al. MRSA carriage in healthcare personnel in contact with farm animals. *J Hosp Infect* 2008;70:186–90.

21. van Rijen MM, Van Keulen PH, Kluytmans JA. Increase in a Dutch hospital of methicillin-resistant Staphylococcus aureus related to animal farming. *Clin Infect Dis* 2008;46:261–63.

22. Nunan C, Young R. MRSA in Farm Animals and Meat: A New Threat to Human Health: The Soil Association, June, 2007.

23. ST398 MRSA infections in Scotland. 2008. (Accessed July 6, 2008, at http://www.documents.hps.scot.nhs.uk/ewr/pdf2008/0823.pdf.)

24. Weese JS, Archambault M, Willey BM, et al. Methicillin-resistant Staphylococcus aureus in horses and horse personnel, 2000–2002. *Emerg Infect Dis* 2005;11:430–35.

25. Interview with J. Scott Weese, DVM, April 29, 2008.

26. Crisostomo, 2001.

27. Khanna T, Friendship R, Dewey C, Weese JS. Methicillin resistant Staphylococcus aureus colonization in pigs and pig farmers. *Vet Microbiol* 2008;128:298–303.

28. Global Agriculture Information Network. *Canada Livestock and Products Livestock Annual 2008*. Washington, DC: USDA Foreign Agricultural Service; 2008 September 11.

29. National Agricultural Statistics Service. *Quarterly Hogs and Pigs*. December 30, 2008. Washington, DC: USDA Agricultural Statistics Board; 2008 December 30.

30. Interview with Tara C. Smith, PhD, Abby L. Harper, MS, and Mike J. Male, DVM, July 11, 2008.

31. Smith TC, Male MJ, Harper AL, Kroeger JS, Tinkler GP, et al. 2009 Methicillin-Resistant Staphylococcus aureus (MRSA) Strain ST398 Is Present in Midwestern U.S. Swine and Swine Workers. PLoS ONE 4(1):e4258. doi:10.1371/journal.pone.0004258, plosone.org/article/info%3Adoi%2F10.1371%2Fjournal.pone.0004258, accessed Jan. 23, 2009.

32. Diep, *Lancet*, 2006.

33. Mellon M, Benbrook C, Benbrook KL. Hogging It! Estimates of Antimicrobiol Abuse in Livestock: Union of Concerned Scientists, January 2001.

34. Stokstad EL, Jukes TH. Effect of various levels of vitamin B12 upon growth response produced by aureomycin in chicks. *Proc Soc Exp Biol Med* 1951;76:73–76.

35. Levy SB, Fitzgerald GG, Macone AB. Changes in intestinal flora of farm personnel after introduction of a tetracycline-supplemented feed on a farm. *N Engl J Med* 1976;295:583–88.

36. The complex politics of avoparcin's cancellation in the European Union are explored in Michael Schnayerson and Mark J. Plotkin's *The Killers Within: The Deadly Rise of Drug-Resistant Bacteria* (Boston: Little, Brown, 2002).

37. Smith KE, Besser JM, Hedberg CW, et al. Quinolone-resistant Campylobacter jejuni infections in Minnesota, 1992–1998. Investigation Team. *N Engl J Med* 1999;340:1525–32.

38. Food and Drug Administration. Abbott Laboratories' Sarafloxacin for Poultry;

Withdrawal of Approval of NADAs. Docket No. 01N-0170. *Federal Register* April 30; 2001:21400–401.

39. Food and Drug Administration. Animal Drugs, Feeds, and Related Products; Enrofloxacin for Poultry; Withdrawal of Approval of New Animal Drug Application. Docket No. 2000N-1571. *Federal Register* August 1; 2005:44048–49.

40. Environmental Protection Agency. National Pollutant Discharge Elimination System Permit Regulation and Effluent Limitation Guidelines and Standards for Concentrated Animal Feeding Operations (CAFOs). 2003:7176–274.

41. Government Accountability Office. Concentrated Animal Feeding Operations: EPA Needs More Information and a Clearly Defined Strategy to Protect Air and Water Quality; 2008 September 24.

42. Union of Concerned Scientists. CAFOs uncovered: The untold cost of confined animal feeding operations. (Accessed July 1, 2008, at http://www.ucsusa.org/food_and_agriculture/science_and_impacts/impacts_industrial_agriculture/cafos-uncovered .html.)

43. National Agricultural Statistical Service. 2002 Census of Agriculture: US Department of Agriculture; 2002.

44. Storteboom HN, Kim SC, Doesken KC, Carlson KH, Davis JG, Pruden A. Response of antibiotics and resistance genes to high-intensity and low-intensity manure management. *J Environ Qual* 2007;36:1695–703.

45. Cleaning up hog waste in North Carolina. 2007. (Accessed January 9, 2009, at http://www.edf.org/page.cfm?tagID=68.)

46. Koike S, Krapac IG, Oliver HD, et al. Monitoring and source tracking of tetracycline resistance genes in lagoons and groundwater adjacent to swine production facilities over a 3-year period. *Appl Environ Microbiol* 2007;73:4813–23.

47. Dolliver H, Kumar K, Gupta S. Sulfamethazine uptake by plants from manure-amended soil. *J Environ Qual* 2007;36:1224–30.

48. Gibbs SG, Green CF, Tarwater PM, Mota LC, Mena KD, Scarpino PV. Isolation of antibiotic-resistant bacteria from the air plume downwind of a swine confined or concentrated animal feeding operation. *Environ Health Perspect* 2006;114:1032–37.

49. Graham JP, Price LB, Evans SL, Graczyk TK, Silbergeld EK. Antibiotic resistant enterococci and staphylococci isolated from flies collected near confined poultry feeding operations. *Sci Total Environ* 2009;407:2701–10.

50. Price LB, Graham JP, Lackey LG, Roess A, Vailes R, Silbergeld E. Elevated risk of carrying gentamicin-resistant Escherichia coli among U.S. poultry workers. *Environ Health Perspect* 2007;115:1738–42.

51. Kluytmans J, van Leeuwen W, Goessens W, et al. Food-initiated outbreak of methicillin-resistant Staphylococcus aureus analyzed by pheno- and genotyping. *J Clin Microbiol* 1995;33:1121–28.

52. van Loo IH, Diederen BM, Savelkoul PH, et al. Methicillin-resistant Staphylococcus aureus in meat products, the Netherlands. *Emerg Infect Dis* 2007;13:1753–55.

53. Weese JS. Methicillin-resistant Staphylococcus aureus in animals and links to human disease. *International Conference on Emerging Infectious Diseases*, 2008 March 19; Atlanta, GA.

54. Interview with J. Scott Weese, DVM, March 19, 2008.

55. Letter to Collin C. Peterson, Chairman, Committee on Agriculture, House of Representatives. 2008. (Accessed at http://nppc.org/CDCLetter.pdf.)

56. Shields D, Mathews K. *Interstate Livestock Movements.* Washington, DC: Economic Research Service, US Department of Agriculture; 2003 June.
57. Dumortier C, Taylor B, Sanchez JE, et al. Evidence of S. aureus transmission between the USA and the Dominican Republic. *48th Annual Interscience Conference on Antimicrobial Agents and Chemotherapy and the Infectious Diseases Society of America 46th Annual Meeting*, 2008 October 25–28; Washington, D.C.

11. SOAP, THE WONDER CURE

1. Discussion of Novant Health's programs is based on observing Patty Deltry, RN, on July 29, 2008; and on interviews in Orlando and Charlotte with Diana Best, RN (December 11, 2007), Suzie Rakyta, RN, Stephen L. Wallenhaupt, MD, and Thomas N. Zweng, MD (December 11, 2007 and July 29, 2008), and Sandy Cox, RN, James Lederer, Jr., MD, and Paul Wiles, president and CEO (July 29, 2008).
2. National Nosocomial Infections Surveillance (NNIS) System Report, data summary from January 1992 to June 2002, issued August 2002. *Am J Infect Control* 2002;30:458–75.
3. In addition to the articles cited, this section is based on interviews with Dr. Barry M. Farr (March 13, 2008) and Dr. William R. Jarvis (February 1, 2008), and on materials from the Problem Pathogen Partnership that are archived on the Internet at http:// www.healthsystem.virginia.edu/Internet/ppp/.
4. Richet HM, Benbachir M, Brown DE, et al. Are there regional variations in the diagnosis, surveillance, and control of methicillin-resistant Staphylococcus aureus? *Infect Control Hosp Epidemiol* 2003;24:334–41.
5. Fazal BA, Telzak EE, Blum S, Turett GS, Petersen-Fitzpatrick FE, Lorian V. Trends in the prevalence of methicillin-resistant Staphylococcus aureus associated with discontinuation of an isolation policy. *Infect Control Hosp Epidemiol* 1996;17:372–74.
6. Barrett SP, Mummery RV, Chattopadhyay B. Trying to control MRSA causes more problems than it solves. *J Hosp Infect* 1998;39:85–93.
7. Jarvis WR, Ostrowsky B. Dinosaurs, methicillin-resistant Staphylococcus aureus, and infection control personnel: survival through translating science into prevention. *Infect Control Hosp Epidemiol* 2003;24:392–96.
8. Boyce JM, Potter-Bynoe G, Chenevert C, King T. Environmental contamination due to methicillin-resistant Staphylococcus aureus: possible infection control implications. *Infect Control Hosp Epidemiol* 1997;18:622–27.
9. Albert RK, Condie F. Hand-washing patterns in medical intensive-care units. *N Engl J Med* 1981;304:1465–66.
10. Donowitz LG. Handwashing technique in a pediatric intensive care unit. *Am J Dis Child* 1987;141:683–85.
11. Meengs MR, Giles BK, Chisholm CD, Cordell WH, Nelson DR. Hand washing frequency in an emergency department. *Ann Emerg Med* 1994;23:1307–12.
12. Monnet DL. Methicillin-resistant Staphylococcus aureus and its relationship to antimicrobial use: possible implications for control. *Infect Control Hosp Epidemiol* 1998;19:552–59.
13. Peacock, 1980.
14. Thompson, 1982.

15. Shanson DC, Kensit JC, Duke R. Outbreak of hospital infection with a strain of Staphylococcus aureus resistant to gentamicin and methicillin. *Lancet* 1976;2:1347–48.
16. Jernigan, 1995.
17. Jernigan JA, Titus MG, Groschel DH, Getchell-White S, Farr BM. Effectiveness of contact isolation during a hospital outbreak of methicillin-resistant Staphylococcus aureus. *Am J Epidemiol* 1996;143:496–504.
18. Salgado CD, Farr BM. What proportion of hospital patients colonized with methicillin-resistant Staphylococcus aureus are identified by clinical microbiological cultures? *Infect Control Hosp Epidemiol* 2006;27:116–21.
19. The Problem Pathogen Partnership. 2000. (Accessed March 1, 2008, at http://www.healthsystem.virginia.edu/internet/ppp/home.cfm.)
20. Muto CA, Jernigan JA, Ostrowsky BE, et al. SHEA guideline for preventing nosocomial transmission of multidrug-resistant strains of Staphylococcus aureus and enterococcus. *Infect Control Hosp Epidemiol* 2003;24:362–86.
21. Rakyta S, Zweng T, 2007. A comprehensive, system-wide program to improve hand hygene compliance within a healthcare system. *19th National Forum on Quality Improvement in Healthcare*, 2007 December 9–12; Orlando, FL: Institute for Healthcare Improvement.
22. Voss A, Widmer AF. No time for handwashing!? Handwashing versus alcoholic rub: can we afford 100% compliance? *Infect Control Hosp Epidemiol* 1997;18:205–8.
23. Sunenshine RH, Liedtke LA, Fridkin SK, Strausbaugh LJ. Management of inpatients colonized or infected with antimicrobial-resistant bacteria in hospitals in the United States. *Infect Control Hosp Epidemiol* 2005;26:138–43.
24. Diekema DJ, Edmond MB. Look before you leap: active surveillance for multidrug-resistant organisms. *Clin Infect Dis* 2007;44:1101–7.
25. Harbath S, Fankhauser C, Schrenzel J, et al. Universal screening for methicillin-resistant Staphylococcus aureus at hospital admission and nosocomial infection in surgical patients. *JAMA* 2008;299:1149–57.
26. Wenzel RP, Bearman G, Edmond MB. Screening for MRSA: a flawed hospital infection control intervention. *Infect Control Hosp Epidemiol* 2008;29:1012–18.
27. Aboelela SW, Saiman L, Stone P, Lowy FD, Quiros D, Larson E. Effectiveness of barrier precautions and surveillance cultures to control transmission of multi-drug-resistant organisms: a systematic review of the literature. *Am J Infect Control* 2006;34:484–94.
28. McGinigle KL, Gourlay ML, Buchanan IB. The use of active surveillance cultures in adult intensive care units to reduce methicillin-resistant Staphylococcus aureus-related morbidity, mortality, and costs: a systematic review. *Clin Infect Dis* 2008;46:1717–25.
29. Clancy M, Graepler A, Wilson M, Douglas I, Johnson J, Price CS. Active screening in high-risk units is an effective and cost-avoidant method to reduce the rate of methicillin-resistant Staphylococcus aureus infection in the hospital. *Infect Control Hosp Epidemiol* 2006;27:1009–17.
30. Espinoza C, Fisher V, Jean W, et al. A pilot study of active surveillance for MRSA by PCR on admission to ICUs at a tertiary care center. K-3367. *48th Annual Interscience Conference on Antimicrobial Agents and Chemotherapy and the Infectious Diseases Society of America 46th Annual Meeting*, October 25–28, 2008, Washington, D.C.
31. Standiford HC, Hebden J, Algire M, Venezia RA, Johnson JK. Early impact of an extensive methicillin-resistant Staphylococcus aureus (MRSA) eradication program

at an academic medical center on hospital-acquired MRSA bacteremia (HA-MRSA BSI). K-459. *47th Annual Interscience Conference on Antimicrobial Agents and Chemotherapy*, September 17–20, 2007, Chicago.

32. Bowler WA, Bresnahan J, Bradfish A. An integrated approach to MRSA control in a rural, regional-referral, health care setting. 28. *18th Annual Scientific Meeting of the Society for Healthcare Epidemiology of America*, April 5–8, 2008, Orlando, FL.

33. Huang SS, Yokoe DS, Hinrichsen VL, et al. Impact of routine intensive care unit surveillance cultures and resultant barrier precautions on hospital-wide methicillin-resistant Staphylococcus aureus bacteremia. *Clin Infect Dis* 2006;43:971–78.

34. Farr BM, Jarvis WR. Methicillin-resistant Staphylococcus aureus: misinterpretation and misrepresentation of active detection and isolation. *Clin Infect Dis* 2008;47:1238–39; author reply 9–40.

35. Management of multidrug-resistant organisms in healthcare settings, 2006. 2006. (Accessed at http://www.cdc.gov/ncidod/dhqp/pdf/ar/mdroGuideline2006.pdf.)

36. Health-care-associated infections in hospitals: Leadership needed from HHS to prioritize prevention practices and improve data on these infections 2008. (Accessed April 16, 2008, at www.gao.gov/cgi-bin/getrpt?GAO-08-283.)

37. This discussion of the growing embrace of active surveillance cultures is based in part on interviews with Betsy McCaughey, PhD, on March 10, 2008, and with Joseph McCannon, Frances Griffin, and Dr. Don Goldmann of the Institute for Healthcare Improvement on December 12, 2007.

38. Robicsek A, Beaumont JL, Paule SM, et al. Universal surveillance for methicillin-resistant Staphylococcus aureus in 3 affiliated hospitals. *Ann Intern Med* 2008;148:409–18.

39. Peterson LR, Hacek DM, Robicsek A. 5 Million Lives Campaign. Case study: an MRSA intervention at Evanston Northwestern Healthcare. *Jt Comm J Qual Patient Saf* 2007;33:732–38.

40. Methicillin-resistant Staphylococcus aureus (MRSA) initiative. Department of Veterans Affairs, 2007. (Accessed December 28, 2007, at http://www1.va.gov/vhapublications/ViewPublication.asp?pub_ID=1525.)

41. The description of Jeanine Thomas's experience is based on interviews with her on July 7, 2006; January 31, 2008; and February 20, 2009. It is also based on interviews with advocates Michael Bennett (November 15, 2007); Carole Moss (February 20 and March 28, 2008); Victoria Nahum (January 3, 2008); and Lisa McGiffert (April 22, 2008).

42. Berens MJ. Infection epidemic carves deadly path across America. *Chicago Tribune* 2002 July 21.

43. MRSA in Illinois: descriptive analysis of hospital discharge data, 2002–2006. 2008. (Accessed February 22, 2009, at http://www.idph.state.il.us/health/infect/MRSA_Data_02–06.pdf.)

44. An up-to-date list of laws and reports is maintained by the Stop Hospital Infections project of Consumers Union at http://www.stophospitalinfections.org.

45. Evans HL, Shaffer MM, Hughes MG, et al. Contact isolation in surgical patients: a barrier to care? *Surgery* 2003;134:180–88.

46. Saint S, Higgins LA, Nallamothu BK, Chenoweth C. Do physicians examine patients in contact isolation less frequently? A brief report. *Am J Infect Control* 2003;31:354–56.

47. Stelfox HT, Bates DW, Redelmeier DA. Safety of patients isolated for infection control. *JAMA* 2003;290:1899–905.

48. Catalano G, Houston SH, Catalano MC, et al. Anxiety and depression in hospitalized patients in resistant organism isolation. *South Med J* 2003;96:141–45.
49. Tarzi S, Kennedy P, Stone S, Evans M. Methicillin-resistant Staphylococcus aureus: psychological impact of hospitalization and isolation in an older adult population. *J Hosp Infect* 2001;49:250–54.
50. This section is based on interviews with Molly Donahue on January 2, 2009; with Molly Donahue and Jane McGuinn in Leland, IL on February 20, 2009; and on discussions with Jane McGuinn's granddaughter Charlotte McGuinn Freeman.

12. THE END OF ANTIBIOTICS

1. Fletcher L. Cubist highlights FDA's antibiotic resistance. *Nat Rev Drug Discov* 2002;1:206–7.
2. Cubist Pharmaceuticals announces results from first phase III Cidecin(R) community-acquired pneumonia trial. 2002. (Accessed January 16, 2002, at http://www1.snl.com/Cache/c1001129725.html.)
3. In addition to the works cited, the sections on the development of daptomycin are based on interviews with personnel from Cubist Pharmaceuticals: Dr. Barry Eisenstein (May 21, June 2, and June 3, 2008); Jared Silverman, PhD, and Grace Thorne, PhD (May 21, 2008); and Richard Baltz (June 5, 2008); and on an unpublished manuscript by Eisenstein, Baltz, and Fredrick B. Oleson Jr. Dr. Kevin Tally, son of Dr. Frank Tally, was interviewed on November 24, 2008.
4. Dubos's and Waksman's discoveries are discussed in "From Microbes to Medicine," by Rollin D. Hotchkiss, in *Launching the Antibiotic Era* (Zanvil Cohn and Carol Moberg, Rockefeller University Press, 1990), and Stuart Levy's *The Antibiotic Paradox* (Perseus Publishing, 2002).
5. Griffith RS. Introduction to vancomycin. *Rev Infect Dis* 1981;3 suppl:S200–204.
6. Daptomycin's origin on Mount Ararat is common knowledge among former Lilly Research Laboratory workers and is recorded in taxonomy databases. Beyond that single fact, however, no historical material on the original identification of daptomycin remains in the company files, according to Lilly corporate archivist Lisa E. Bayne (email, February 9, 2009).
7. Levine, 2006.
8. Before the 1962 passage of the Kefauver-Harris Amendment to the 1938 Food, Drug and Cosmetic Act, drug manufacturers were required to prove only that their compounds were not unsafe. The amendment added a proof-of-efficacy requirement that in time gave rise to modern clinical trials.
9. Geraci JE, Heilman FR, Nichols DR, Ross GT, Wellman WE. Some laboratory and clinical experiences with a new antibiotic, vancomycin. *Proc Staff Meet Mayo Clin* 1956;31:564–82.
10. Kirby WM. Vancomycin therapy in severe staphylococcal infections. *Rev Infect Dis* 1981;3 suppl:S236–39.
11. Kirst HA, Thompson DG, Nicas TI. Historical yearly usage of vancomycin. *Antimicrob Agents Chemother* 1998;42:1303–4.
12. Kollef MH. Limitations of vancomycin in the management of resistant staphylococcal infections. *Clin Infect Dis* 2007;45 Suppl 3:S191–95.
13. Huycke M M, Sahm DF, Gilmore MS. Multiple-drug resistant enterococci: the nature of the problem and an agenda for the future. *Emerg Infect Dis* 1998;4:239–49.

14. Morris JG, Jr., Shay DK, Hebden JN, et al. Enterococci resistant to multiple antimicrobial agents, including vancomycin. Establishment of endemicity in a university medical center. *Ann Intern Med* 1995;123:250–59.

15. Hiramatsu K, Hanaki H, Ino T, Yabuta K, Oguri T, Tenover FC. Methicillin-resistant Staphylococcus aureus clinical strain with reduced vancomycin susceptibility. *J Antimicrob Chemother* 1997;40:135–36.

16. Hiramatsu K, Aritaka N, Hanaki H, et al. Dissemination in Japanese hospitals of strains of Staphylococcus aureus heterogeneously resistant to vancomycin. *Lancet* 1997;350:1670–73.

17. In the United States, MICs are reported in micrograms per milliliter, rather than milligrams per liter as they were in Japan.

18. Centers for Disease Control and Prevention. Staphylococcus aureus resistant to vancomycin—United States, 2002. *MMWR, Morb Mortal Wkly Rep* 2002; 51:565–67.

19. Centers for Disease Control and Prevention. Vancomycin-resistant Staphylococcus aureus—Pennsylvania, 2002. *MMWR, Morb Mortal Wkly Rep* 2002;51:902.

20. Tenover FC. Vancomycin-resistant Staphylococcus aureus: a perfect but geographically limited storm? *Clin Infect Dis* 2008;46:675–77.

21. Finks J, Wells E, Dyke TL, et al. Vancomycin-resistant Staphylococcus aureus, Michigan, USA, 2007. *Emerging Infectious Diseases* 2009: Epub ahead of print.

22. Liu C, Chambers HF. Staphylococcus aureus with heterogeneous resistance to vancomycin: epidemiology, clinical significance, and critical assessment of diagnostic methods. *Antimicrob Agents Chemother* 2003;47:3040–45.

23. Deresinski S. Counterpoint: Vancomycin and Staphylococcus aureus—an antibiotic enters obsolescence. *Clin Infect Dis* 2007;44:1543–48.

24. Debono M, Abbott BJ, Molloy RM, et al. Enzymatic and chemical modifications of lipopeptide antibiotic A21978C: the synthesis and evaluation of daptomycin (LY146032). *J Antibiot* (Tokyo) 1988;41:1093–1105.

25. Cotroneo N, Harris R, Perlmutter N, Beveridge T, Silverman JA. Daptomycin exerts bactericidal activity without lysis of Staphylococcus aureus. *Antimicrob Agents Chemother* 2008;52:2223–35.

26. Projan SJ. Obituary: Francis P. Tally, MD. *Nat Rev Drug Disc* 2006:18.

27. In addition to the works cited, this section benefited from an interview with Dr. John Powers, former lead medical officer for antibiotic development and resistance initiatives of the Food and Drug Administration (July 6, 2007), and from a presentation on "Resistance and Public Policy" he gave at the 2007 National Conference on Antimicrobial Resistance in Bethesda, MD (June 25, 2007).

28. Tillotson GS. Where does novel antibiotics R&D stand among other pharmaceutical products: an industrial perspective? *Expert Rev Anti Infect Ther* 2008;6:551–52.

29. DiMasi JA, Hansen RW, Grabowski HG. The price of innovation: new estimates of drug development costs. *J Health Econ* 2003;22:151–85.

30. Spellberg B, Powers JH, Brass EP, Miller LG, Edwards JE, Jr. Trends in antimicrobial drug development: implications for the future. *Clin Infect Dis* 2004;38:1279–86.

31. Tsiodras, 2001.

32. Kaplan SL. Treatment of community-associated methicillin-resistant Staphylococcus aureus infections. *Pediatr Infect Dis J* 2005;24:457–58.

33. Projan SJ. Why is big pharma getting out of antibacterial drug discovery? *Curr Opin Microbiol* 2003;6:427–30.

34. Powers JH. Development of drugs for antimicrobial-resistant pathogens. *Curr Opin Infect Dis* 2003;16:547–51.
35. Spellberg, 2004.
36. Boucher HW, Talbot GH, Bradley JS, et al. Bad bugs, no drugs: no ESKAPE! An update from the Infectious Diseases Society of America. *Clin Infect Dis* 2009;48:1–12.
37. Malani PN, Rana MM, Banerjee M, Bradley SF. Staphylococcus aureus bloodstream infections: the association between age and mortality and functional status. *J Am Geriatr Soc* 2008;56:1485–89.
38. Bishop E, Melvani S, Howden BP, Charles PG, Grayson ML. Good clinical outcomes but high rates of adverse reactions during linezolid therapy for serious infections: a proposed protocol for monitoring therapy in complex patients. *Antimicrob Agents Chemother* 2006;50:1599–602.
39. Interview with Dr. John Edwards, Jr., Harbor-UCLA Medical Center, Los Angeles, by phone, May 24, 2007.
40. FDA Approves Cubicin, previously Cidecin, an antibiotic developed by Cubist Pharmaceuticals. 2003. (Accessed at http://www1.snl.com/Cache/c1001129680.html.)
41. Arbeit RD, Maki D, Tally FP, Campanaro E, Eisenstein BI. The safety and efficacy of daptomycin for the treatment of complicated skin and skin-structure infections. *Clin Infect Dis* 2004;38:1673–81.
42. Fowler VG, Jr., Boucher HW, Corey GR, et al. Daptomycin versus standard therapy for bacteremia and endocarditis caused by Staphylococcus aureus. *N Engl J Med* 2006;355:653–65.
43. Center for Drug Evaluation and Research. Minutes of the Anti-Infective Drug Advisory Committee, March 6, 2006. In: *Food and Drug Administration*; 2006:1–323.
44. Jacobson LM, Milstone AM, Zenilman J, Carroll KC, Arav-Boger R. Daptomycin therapy failure in an adolescent with methicillin-resistant Staphylococcus aureus bacteremia. *Pediatr Infect Dis J* 2009.
45. In addition to the papers cited, the discussion of restricting antibiotic prescribing is based on a phone interview with Dr. Jeffrey A. Linder on July 25, 2007, and on his June 26, 2007, presentation to the National Conference on Antimicrobial Resistance, "Physicians and Inappropriate Prescribing: What Can Be Done to Change Their Behavior?"
46. Gonzales R, Malone DC, Maselli JH, Sande MA. Excessive antibiotic use for acute respiratory infections in the United States. *Clin Infect Dis* 2001;33:757–62.
47. McCaig LF, Besser RE, Hughes JM. Trends in antimicrobial prescribing rates for children and adolescents. *JAMA* 2002;287:3096–3102.
48. Sharp HJ, Denman D, Puumala S, Leopold DA. Treatment of acute and chronic rhinosinusitis in the United States, 1999–2002. *Arch Otolaryngol Head Neck Surg* 2007;133:260–65.
49. Gonzales R, Corbett KK, Leeman-Castillo BA, et al. The "minimizing antibiotic resistance in Colorado" project: impact of patient education in improving antibiotic use in private office practices. *Health Serv Res* 2005;40:101–16.
50. McCaig, 2002.
51. Perz JF, Craig AS, Coffey CS, et al. Changes in antibiotic prescribing for children after a community-wide campaign. *JAMA* 2002;287:3103–9.
52. Gonzales, 2005.

53. Linder JA. Editorial commentary: antibiotics for treatment of acute respiratory tract infections: decreasing benefit, increasing risk, and the irrelevance of antimicrobial resistance. *Clin Infect Dis* 2008;47:744–46.

54. Johann-Liang R, Pamer C, Governale L, et al. Update on national outpatient utilization of oral microbial drugs, 1992–2006. *Annual Conference on Antimicrobial Resistance*, 2007 June 25–27; Bethesda, MD: National Foundation for Infectious Diseases.

55. Alliance for the Prudent Use of Antibiotics. *Massachusetts Physician Survey: Pilot Survey of Primary Care Physicians in Massachusetts*, 1998. Medford, MA: Tufts University; 1999.

56. Linder JA, Singer DE. Desire for antibiotics and antibiotic prescribing for adults with upper respiratory tract infections. *J Gen Intern Med* 2003;18:795–801.

57. Diagnosis and management of acute otitis media. *Pediatrics* 2004;113:1451–65.

58. Finkelstein JA, Stille CJ, Rifas-Shiman SL, Goldmann D. Watchful waiting for acute otitis media: are parents and physicians ready? *Pediatrics* 2005;115:1466–73.

59. Shlaes DM, Gerding DN, John JF, Jr., et al. Society for Healthcare Epidemiology of America and Infectious Diseases Society of America Joint Committee on the Prevention of Antimicrobial Resistance: guidelines for the prevention of antimicrobial resistance in hospitals. *Clin Infect Dis* 1997;25:584–99.

60. Lawton RM, Fridkin SK, Gaynes RP, McGowan JE, Jr. Practices to improve antimicrobial use at 47 US hospitals: the status of the 1997 SHEA/IDSA position paper recommendations. Society for Healthcare Epidemiology of America/Infectious Diseases Society of America. *Infect Control Hosp Epidemiol* 2000;21:256–59.

61. Gerding DN. Good antimicrobial stewardship in the hospital: fitting, but flagrantly flagging. *Infect Control Hosp Epidemiol* 2000;21:253–55.

62. Pakyz AL, MacDougall C, Oinonen M, Polk RE. Trends in antibacterial use in US academic health centers: 2002 to 2006. *Arch Intern Med* 2008;168:2254–60.

63. Dr. Christoph Lehmann and Dr. Alison Agwu were interviewed by phone on September 8, 2008, and at Johns Hopkins on October 29, 2008.

64. Hayes KA, Lehmann CU. The interactive patient: a multimedia interactive educational tool on the World Wide Web. *MD Comput* 1996;13:330–34.

65. Agwu AL, Lee CK, Jain SK, et al. A World Wide Web–based antimicrobial stewardship program improves efficiency, communication, and user satisfaction and reduces cost in a tertiary care pediatric medical center. *Clin Infect Dis* 2008;47:747–53.

66. Buising KL, Thursky KA, Robertson MB, et al. Electronic antibiotic stewardship—reduced consumption of broad-spectrum antibiotics using a computerized antimicrobial approval system in a hospital setting. *J Antimicrob Chemother* 2008;62:608–16.

67. Evans RS, Pestotnik SL, Classen DC, et al. A computer-assisted management program for antibiotics and other antiinfective agents. *N Engl J Med* 1998;338:232–38.

13. THE EPIDEMICS CONVERGING

1. The description of Diane Lore's experience is based on an interview with her on May 12, 2008; on her medical and pharmacy records; on correspondence between her lawyers and the hospital where she delivered; and on conversations with her between 2003 and 2006, when we simultaneously worked at the *Atlanta Journal-Constitution*.

2. Kalstone C. Methicillin-resistant staphylococcal mastitis. *Am J Obstet Gynecol* 1989;161:120.

3. In addition to the cited articles, this section on the unfolding of community-acquired MRSA in hospitals is based on an interview with Dr. Lisa Saiman on April 21, 2009, and on her paper on the initial outbreak: Saiman L, O'Keefe M, Graham PL, III, et al. Hospital transmission of community-acquired methicillin-resistant Staphylococcus aureus among postpartum women. *Clin Infect Dis* 2003;37(10):1313–19.

4. Moran, Methicillin-resistant . . . infections . . . in the emergency department.

5. Bratu S, Eramo A, Kopec R, et al. Community-associated methicillin-resistant Staphylococcus aureus in hospital nursery and maternity units. *Emerg Infect Dis* 2005;11:808–13.

6. Healy CM, Hulten KG, Palazzi DL, Campbell JR, Baker CJ. Emergence of new strains of methicillin-resistant Staphylococcus aureus in a neonatal intensive care unit. *Clin Infect Dis* 2004;39:1460–66.

7. Zaoutis TE, Toltzis P, Chu J, et al. Clinical and molecular epidemiology of community-acquired methicillin-resistant Staphylococcus aureus infections among children with risk factors for health care-associated infection: 2001–2003. *Pediatr Infect Dis J* 2006;25:343–48.

8. Centers for Disease Control and Prevention. Community-associated methicillin-resistant Staphylococcus aureus infection among healthy newborns—Chicago and Los Angeles County, 2004. *MMWR, Morb Mortal Wkly Rep* 2006;55:329–32.

9. Dehority W, Wang E, Vernon PS, Lee C, Perdreau-Remington F, Bradley J. Community-associated methicillin-resistant Staphylococcus aureus necrotizing fasciitis in a neonate. *Pediatr Infect Dis J* 2006;25:1080–81.

10. Seybold U, Kourbatova EV, Johnson JG, et al. Emergence of community-associated methicillin-resistant Staphylococcus aureus USA300 genotype as a major cause of health care–associated blood stream infections. *Clin Infect Dis* 2006;42:647–56.

11. Gonzalez BE, Rueda AM, Shelburne SA, III, Musher DM, Hamill RJ, Hulten KG. Community-associated strains of methicillin-resistant Staphylococcus aureus as the cause of healthcare-associated infection. *Infect Control Hosp Epidemiol* 2006;27:1051–56.

12. Manian FA, Griesnauer S. Community-associated methicillin-resistant Staphylococcus aureus (MRSA) is replacing traditional health care-associated MRSA strains in surgical-site infections among inpatients. *Clin Infect Dis* 2008;47:434–35.

13. Hardman K, Blake RK, Salgado CD. The impact of USA 300 community-acquired methicillin-resistant Staphylococcus aureus (CA-MRSA) on healthcare-acquired MRSA infections (HA-MRSA-I) 119. *18th Annual Scientific Meeting of the Society for Healthcare Epidemiology of America*; 2008 April 5–8; Orlando, FL.

14. Neofytos D, Kuhn B, Shen S, Hua Zhu X, Jungkind D, Flomenberg P. Emergence of staphylococcal cassette chromosome mec type IV methicillin-resistant Staphylococcus aureus as a cause of ventilator-associated pneumonia. *Infect Control Hosp Epidemiol* 2007;28:1206–9.

15. Maree CL, Daum RS, Boyle-Vavra S, Matayoshi K, Miller LG. Community-associated methicillin-resistant Staphylococcus aureus isolates causing healthcare-associated infections. *Emerg Infect Dis* 2007;13:236–42.

16. Popovich KJ, Weinstein RA, Hota B. Are community-associated methicillin-resistant Staphylococcus aureus (MRSA) strains replacing traditional nosocomial MRSA strains? *Clin Infect Dis* 2008;46:787–94.

NOTES to Pages 204–209

17. Klevens RM, Morrison MA, Fridkin SK, et al. Community-associated methicillin-resistant Staphylococcus aureus and healthcare risk factors. *Emerg Infect Dis* 2006;12:1991–93.
18. David MZ, Glikman D, Crawford SE, et al. What is community-associated methicillin-resistant Staphylococcus aureus? *J Infect Dis* 2008;197:1235–43.
19. Stumpf PG, Flores M, Murillo J. Serious postpartum infection due to MRSA in an asymptomatic carrier: case report and review. *Am J Perinatol* 2008;25:413–15.
20. This section on the necrotizing pneumonia outbreak in University of Chicago's NICU is based on an interview with Dr. John Marcinak (February 19, 2009) and on his paper, "Cluster of necrotizing pneumonia caused by community-associated methicillin-resistant Staphylococcus aureus in very low birth weight infants," presented in October 2008 at the *48th Annual Interscience Conference on Antimicrobial Agents and Chemotherapy and the Infectious Diseases Society of America 46th Annual Meeting.*
21. Gelband H. A shot against MRSA? (Policy Brief 7). Washington DC: Resources for the Future; 2009 March.
22. Centers for Disease Control and Prevention. Progress toward elimination of Haemophilus influenzae type b disease among infants and children—United States, 1987–1993. *MMWR, Morb Mortal Wkly Rep* 1994;43:144–48.
23. Whitney, 2003.
24. Bogaert D, Veenhoven RH, Sluijter M, et al. Molecular epidemiology of pneumococcal colonization in response to pneumococcal conjugate vaccination in children with recurrent acute otitis media. *J Clin Microbiol* 2005;43:74–83.
25. Gelband, 2009.
26. Bancroft, 2007.
27. Lucero CA, Zell ER, Hageman J, Bulens SN, Schaffner W, Fridkin SK. Potential impact of S. aureus vaccine on invasive methicillin-resistant Staphylococcus aureus disease in the United States. 344. *18th Annual Scientific Meeting of the Society for Healthcare Epidemiology of America*, April 5–8, 2008, Orlando, FL.
28. Shinefield HR. Use of a conjugate polysaccharide vaccine in the prevention of invasive staphylococcal disease: is an additional vaccine needed or possible? *Vaccine* 2006;24 Suppl 2:S2-65–69.
29. Schaffer AC, Lee JC. Staphylococcal vaccines and immunotherapies. *Infect Dis Clin North Am* 2009;23:153–71.
30. Shinefield H, Black S, Fattom A, et al. Use of a Staphylococcus aureus conjugate vaccine in patients receiving hemodialysis. *N Engl J Med* 2002;346:491–96.
31. Schaffer, 2009.
32. DeJonge M, Burchfield D, Bloom B, et al. Clinical trial of safety and efficacy of INH-A21 for the prevention of nosocomial staphylococcal bloodstream infection in premature infants. *J Pediatr* 2007;151:260–65, 5 e1.
32. Kuklin NA, Clark DJ, Secore S, et al. A novel Staphylococcus aureus vaccine: iron surface determinant B induces rapid antibody responses in rhesus macaques and specific increased survival in a murine S. aureus sepsis model. *Infect Immun* 2006;74:2215–23.
34. Bubeck Wardenburg J, Schneewind O. Vaccine protection against Staphylococcus aureus pneumonia. *J Exp Med* 2008;205:287–94.

EPILOGUE

1. Mandell LA, Wunderink RG, Anzueto A, et al. Infectious Diseases Society of America/American Thoracic Society consensus guidelines on the management of community-acquired pneumonia in adults. *Clin Infect Dis* 2007;44 Suppl 2:S27–72.
2. Fiore AE, Shay DK, Broder K, et al. Prevention and control of influenza: recommendations of the Advisory Committee on Immunization Practices (ACIP), 2008. *MMWR Recomm Rep* 2008;57:1–60.

INDEX

255

INDEX

antibiotics (*cont.*)
   prison staff's use of, 136, 137, 138
   PVL and, 62
   rate of increase of prescriptions for,
     80–81
   restriction of use of, 159, 162
   Sorensen's use of, 36–39
   staph's defenses against, 6, 8, 17, 24
   Thomas's use of, 169
   as time-dependent, 186
   use of fewer, 188–94
   workings of, 182
   see also beta-lactams; daptomycin;
     penicillin; vancomycin
antibiotic stewardship, 190–93
anticlotting medications, 108
antimicrobials, in food animals, 150–51
appetite, 2, 115
Ararat, Mount, 177, 248n
Arkansas, 153
armpits, 41, 46, 83, 127
Army, U.S., 15
Askenasy, Kepa, 64–66, 75, 76–77
aspirin, 172–73
Aspirus Wausau Hospital, 165–66
asthma, 102, 106
athletics, *see* sports teams
Atlanta, Ga., 120, 194–97
   hospitals in, 41, 68, 73, 195, 203
Augmentin, 213
Australia, 24, 43, 50, 62, 97–98, 185
autopsies, 16, 54–55, 61, 101
Avelox, 81, 82
Aventis, 184
avian influenza, 90, 107
avoparcin, 152, 243n
azithromycin, 81, 103, 189, 196
aztreonam, 214

Baby (elephant), 91–93
bacteremia, 43, 47, 52, 132, 138, 203, 215
   daptomycin and, 182, 187–88
bacteria:
   acquired by newborns, 27
   DNA alteration in, 56
   synthetic penicillin and, 34
   vancomycin and, 12–13
   see also specific bacteria

bacterial interference, 30
Bactrim, 78, 214
Bactroban, 82, 86, 90
Baltimore, Md., 58, 73, 98, 179
   jails in, 131, 134
Baltimore VA Medical Center, 69
Baltz, Richard, 176, 183–84, 186
Bancroft, Elizabeth, 130
Barcelona, 50, 77
bathrooms, school, 120–21
baths, hot, 112
Baylor College of Medicine, 59–60
Beecham Research Laboratories, 33, 230n
beef, 155
Bell, James (Abdul-Muhaymin Nuruddin),
   131–33
Belzley, Gregory, 134–35
Bennett, Michael, 170
Benson, Gary, 136–37, 138
beta-lactam ring, 34, 51, 52, 213–14
beta-lactams, 8, 10, 15, 16, 53, 56, 57, 67,
   74, 81, 113, 145, 178, 182, 197, 206,
   213
   Keflex, 81, 82
   level of drug in the blood and, 186
   nafcillin, 187, 213
   Nick Johnson's use of, 112
   Rocephin (ceftriaxone), 176, 214
   vancomycin compared with, 178
Bielski, Robert, 12, 14, 17
bioterrorism, 107
biotyping, 225n
bleach, 86, 122
blood, 52, 59, 99, 154, 165
   daptomycin and, 183
   dialysis graft and, 131–32
   of Lore, 201, 202
   of Nick Johnson, 112–13, 115
   oxygen in, 106, 112, 206
blood clots, 54, 59–60, 109, 207
   pulmonary embolism, 38
blood pressure, 17, 54, 102, 108, 112
   of Tony Love, 2, 3, 4, 14
blood-pressure cuffs, 131
blood vessels, 3, 106, 108
boils, 5, 24, 68, 73, 81, 82, 89, 121, 138
   in football squads, 116–17, 119
   in prisoners, 127, 129, 130, 135

256

Europe, 141, 148, 155
  see also specific countries and cities
European Union, 147, 152, 243n
Evanston Northwestern Healthcare, 167
eye infections, 60
eyelids, 64, 82

Fairfield, Conn., 116–17
Farnham, MacDonald, 140–41
Farr, Barry M., 160–62, 165, 166
fats, body, 187
fertilizer, 153
fever, 15, 21, 24, 47, 53, 55, 59, 97, 98, 103,
      112, 118, 136, 144, 203
  of Askenasy, 65, 66
  of Bell, 132
  of Carlos Don IV, 105
  diphtheria and, 29
  of Isabella Logan, 85
  of Keller, 128
  of Nick Johnson, 112, 115
  of Thomas, 168
  of Tony Love, 2, 3, 11, 12, 14
Fleming, Alexander, 22, 23
flesh-eating disease (necrotizing fasciitis),
      69, 74, 96, 118, 203
Florey, Howard, 22
flu, flu patients, 25, 100–104, 107–8, 136,
      212
  avian, 90, 107
flucloxacillin, 213
fluoroquinolone-resistant Campylobacter
      jejuni, 152
fluoroquinolones, 65, 75, 81, 182, 189,
      214
  Avelox, 81, 82
flu shots, 101, 104, 212
Folsom State Prison, 135–39
Food and Drug Administration, U.S.
      (FDA), 13–14, 25, 152, 187, 214,
      248n
  Anti-Infective Drugs Advisory
      Committee, 187–88
  drug licensing and, 176, 178, 184, 188
food supply, 139, 154–56
football, 111, 113, 114, 116–20, 123, 135
"Four Cs," 130
FPR3757, 74, 75

France, 60–61, 62
  pneumonia in, 97–98, 102

Gansey, Mike, 119
gay men, xi, 77, 118
Geldrop, 146–47
gene-sequencing technique, 67
genetics, genes, 49, 55–58, 61–62, 67, 75,
      180, 181
genome, 61
gentamicin, 78, 214
George, Kenny, 119
Georgia, 103, 123, 129, 153, 170
  hospitals in, 41, 68, 73, 195, 203
Gerberding, Julie, 155
GlaxoSmithKline, 230n
gloves, 158, 159, 173, 201
glycopeptide, 12, 177, 214
Godfrey, Mary, 26
Gooden, Drew, 119
Government Accountability Office (GAO),
      152–53, 166–67
Grady Memorial Hospital, 68, 203
gramicidin, 177
Gram-negatives, 182
Gram-positives, 182
Grand Rounds, 198–99
Great Britain, 40, 62, 66, 84
groin, 41, 46, 65, 86, 136, 138
growth promotion, 151, 152
Grubb, Clarence, 132
guidelines, 212
  HICPAC, 166–67
  SHEA, 162, 170
gum disease, 132
gut, 37, 179, 185–86

H1N1 influenza pandemic, 100, 101
Haemophilus influenzae type B (Hib), 208
Hageman, Jeffrey, 102, 103, 104
hair, body, shaving of, 117
hands, 2, 4
  washing of, 158–66, 201
hantavirus, 107
Harbor-UCLA Medical Center, 69, 185,
      204
Harborview Medical Center, 45–46, 159
Harper, Abby, 141

INDEX

lincosamides, 214
Lindenmayer, Joann, 114
linezolid (Zyvox), 13–14, 47, 75, 81, 83,
    184, 214
    in Australia, 185
    Lore's use of, 200, 201, 202
lipoglycopeptides, 214
lipopeptide, 182, 214
liver, 4, 16, 103
Liverpool, University of, Small Animal
    Hospital of, 89
Logan, Brian, 84–86, 211
Logan, Ian, 211–12
Logan, Isabella, 84–86, 205
Logan, Mollie, 84–86, 211–12
London, 90, 97–98
    Central Middlesex Hospital in, 40–41
Lore, Diane, 195–97, 200–202, 212
Los Angeles, Calif., 69, 185, 203, 204
    HIV cases in, 15, 77
Los Angeles County, 68
    jail in, 118, 129–30
Los Angeles County Department of Health
    Services, 118
Los Angeles Department of Public Health,
    xi, 130
Los Angeles Howard Jones Field, 118
Los Angeles Sheriff's Department, 129
Louisiana, 103, 131
Love, Clarissa, 1, 3, 4, 11, 12, 14, 17, 211
Love, Tony, 1–5, 11–14, 17, 211
Lucasville, Ohio, 131–32
Lungile (elephant), 92
lungs, 6, 12, 17, 53, 54, 60, 61, 205
    abscesses in, 16, 43
    artificial, 99
    of Bryce Smith, 96
    cancer of, 179
    of Carlos Don IV, 106–9
    daptomycin and, 186, 187
    infection in, 43, 154
    of Nick Johnson, 112, 113, 115
    pneumonia and, 96–99, 101, 102, 104
    of premature babies, 206, 207
    pulmonary surfactants and, 187
    vancomycin and, 178
Lycoming College, 119–20
lymph nodes, 112, 136

Macario, Everly, 53–55, 211
McCaughey, Betsy, 167
McGuinn, Jane, 171–74
Mackin property, 172, 173, 174
McNees, Steve, 81–83, 211
macrolides, 74, 75, 81, 182, 189, 200,
    214
    see also azithromycin; erythromycin
Maimonides Medical Center, 200
Male, Mike, 141, 149–50
manure, from CAFOs, 153
Martin, Benjamin, 128, 212
Maryland, 153–54, 170
Maryland, University of, 133–34, 165
Massachusetts, 79, 152, 190
mastitis (teat infection), 143, 144, 196–97,
    198
Maxon, Thelma, 23
mecA, 55–56, 58, 61, 179
Memorial Hermann Hospital, 43, 165
Memphis, Tenn., 44, 58, 60
meningitis, 28, 55, 178, 208
mental hospitals, 171
Merck & Co., 209
meropenem, 214
methadone, 100
methicillin, 6, 10, 12, 40, 178, 213
    demise of, 230*n*
    penicillin compared with, 51
methicillin-resistant *Staphylococcus*
    aureus, see MRSA
Miami Dolphins, 114
MIC, *see* minimum inhibitory
    concentration
MIC creep, 181
Michigan, 149, 180, 181
minimum inhibitory concentration
    (MIC), 180, 181, 188, 249*n*
Minneapolis, Minn., 15, 16, 58, 67
Minnesota, 15, 16, 53, 58, 67, 68, 70–71,
    198
    CAFO waste in, 153
    community MRSA in, 73, 128
    pigs in, 140–41, 149
Minnesota Department of Health, 152
Mississippi, 11, 15, 68, 129
MLST, *see* multilocus sequence typing
molecular analysis, 30, 50, 66

INDEX

Tennessee Department of Health, 79
testicles, 126, 127, 137
tetracyclines, 24, 26, 45, 56, 75, 78, 81,
   182, 214
  in chicken feed, 152
  ST398's resistance to, 145, 150
  in U.S. vs. Netherlands, 145
Texas, 15, 33, 83
  CAFO research in, 153
  county jail in, 131
  football as business in, 116
  hospitals in, 26–27, 43, 58, 59–60,
   112–15, 165, 202–3
Texas, University of (Austin), 113
Texas Children's Hospital, 59–60, 62,
   112–15, 202–3
Texas Department of State Health
   Services, 113, 116
Thiel College, 114
Thomas, Jeanine, 168–71
Thomas Jefferson University Hospital,
   203
throat, 28–30, 83, 100, 101, 155
  infection of, 53
  sore, 29, 30, 103, 189
thrush, 39
tigecycline, 183
time-dependent, 186
towels, 118, 119, 120
toxic shock, 5, 66, 75, 132, 205, 208
tracheostomy, 43
transduction, 56, 230n
transformation, 56, 230n
transplant rejection, 52
trimethoprim-sulfamethoxazole, 78, 81,
   83, 200, 206, 211
  see also Septra
Trojans, 117–18
tuberculosis, 176
Tullis, David, 83
turf burns, 117
Tylenol, 94, 103, 111
type G, 67

UCSF, see California, University of, at
   San Francisco
ultrasound guided needles, 115
Union of Concerned Scientists, 150–51

United Kingdom, 34, 147
  see also Great Britain
United States:
  CAFOs in, 152–54
  epidemic of antibiotic-resistant staph
   in, 24–27
  first MRSA cases in, 43, 50, 58, 159
  naming schemes in, 67–68
  Netherlands compared with, 142,
   145, 150
  penicillin manufacture and sale in,
   23
  pigs in, 140–41, 146, 149, 152, 153,
   155–56
  poultry growth promoters in, 152
  start of methicillin sales in, 34
  see also specific topics
University Hospital, 154–56
urinary tract infections, 43, 77, 198
urologists, 137–38
USA100, 67, 78, 205
USA200, 67
USA300, 67–70, 74–78, 80, 83, 86,
   181–82, 203–7
  in animals, 87–88, 89, 93
  diversification of, 78
  football squads and, 116
  pneumonia and, 102
  in San Francisco, 71–72, 74, 78, 131
USA300-0114, 69, 74, 78
USA400, 67–68, 102, 134, 198, 199
USA500, 67, 148
USA600, 67
USA700, 68
USA800, 67

vaccine:
  flu, 101, 104, 208
  MRSA, 207–10
  pneumococcus, 103, 208, 209
vagina, 83, 205, 212
valley fever, 107
VA Medical Center, 79
vanA, 181
vancomycin, 3, 8, 12–13, 46, 47, 85, 96,
   107, 137, 138, 177–82, 187, 201–2,
   206, 211, 214
  avoparcin compared with, 152

269

ABOUT THE AUTHOR

Maryn McKenna is an award-winning science journalist and author of *Beating Back the Devil: On the Front Lines with the Disease Detectives of the Epidemic Intelligence Service* (named one of the top ten science books of 2004 by Amazon). She writes for national magazines and medical journals; previously, she was a newspaper reporter and a staff member at the Center for Infectious Disease Research and Policy at the University of Minnesota. She has been a fellow with the Dart Center for Journalism and Trauma at Columbia University, the Henry J. Kaiser Family Foundation, and the Knight-Wallace Program of the University of Michigan. She is a graduate of Georgetown University and the Medill School of Journalism at Northwestern University. She lives in Minneapolis and Atlanta.